U.S.NRC

United States Nuclear Regulatory Commission

Protecting People and the Environment

NUREG-1909

I0482773

Background, Status, and Issues Related to the Regulation of Advanced Spent Nuclear Fuel Recycle Facilities

ACNW&M White Paper

Advisory Committee on Nuclear Waste and Materials

U.S.NRC

United States Nuclear Regulatory Commission

Protecting People and the Environment

NUREG-1909

Background, Status, and Issues Related to the Regulation of Advanced Spent Nuclear Fuel Recycle Facilities

ACNW&M White Paper

Manuscript Completed: May 2008
Date Published: June 2008

Prepared by
A.G. Croff, R.G. Wymer, L.L. Tavlarides, J.H. Flack, H.G. Larson

Advisory Committee on Nuclear Waste and Materials

THIS PAGE WAS LEFT BLANK INTENTIONALLY

ABSTRACT

In February 2006, the Commission directed the Advisory Committee on Nuclear Waste and Materials (ACNW&M) to remain abreast of developments in the area of spent nuclear fuel reprocessing, and to be ready to provide advice should the need arise. A white paper was prepared in response to that direction and focuses on three major areas: (1) historical approaches to development, design, and operation of spent nuclear fuel recycle facilities, (2) recent advances in spent nuclear fuel recycle technologies, and (3) technical and regulatory issues that will need to be addressed if advanced spent nuclear fuel recycle is to be implemented. This white paper was sent to the Commission by the ACNW&M as an attachment to a letter dated October 11, 2007 (ML072840119). In addition to being useful to the ACNW&M in advising the Commission, the authors believe that the white paper could be useful to a broad audience, including the NRC staff, the U.S. Department of Energy and its contractors, and other organizations interested in understanding the nuclear fuel cycle.

******

THIS PAGE WAS LEFT BLANK INTENTIONALLY

TABLE OF CONTENTS

FIGURES

TABLES

EXECUTIVE SUMMARY

INTRODUCTION

The United States currently has 104 operating commercial nuclear power reactors that produce about 2100 metric tons of initial heavy metal (MTIHM) of spent nuclear fuel (SNF) each year. The U.S. Department of Energy (DOE) estimates that the congressionally mandated capacity limit of 70,000 MT of heavy metal equivalent imposed on the proposed Yucca Mountain repository will be committed to accumulated spent commercial fuel and other DOE wastes by about 2010. The SNF from existing and future nuclear power reactors in the United States poses the following challenges:

- the desire to create additional disposal capacity without creating additional repositories

- the potential to increase utilization of the fissile and fertile material that constitute about 1 percent and 95 percent of the SNF, respectively, by recovering and recycling them[1]

- avoiding the proliferation risk from production and use of a pure plutonium stream in recycle

- reducing the long-term repository risk from key radionuclides in SNF such as ^{99}Tc, ^{129}I, and ^{237}Np

To address these challenges, DOE is proposing to reprocess SNF, primarily from light-water reactors (LWRs) in the foreseeable future; reuse the recovered uranium directly or through reenrichment; reuse the plutonium by making it into new reactor fuel (refabrication); destroy actinides that dominate repository risk by refabricating them into fuel or targets and irradiating the actinides in a nuclear reactor; and incorporating radionuclides that cannot be readily destroyed by irradiation into tailored waste forms. To address proliferation concerns, DOE proposes to reprocess the SNF using new approaches that do not produce a separated plutonium stream.

The current DOE program for implementing SNF recycle contemplates building three facilities— an integrated nuclear fuel recycle facility; an advanced reactor for irradiating neptunium, plutonium, americium, and curium; and an advanced fuel cycle research facility to develop recycle technology. The first two of these are likely to be licensed by the U.S. Nuclear Regulatory Commission (NRC).

Fuel recycle has the potential to require changes in the NRC's existing regulatory framework and expertise which are now structured to license LWRs and their associated once-through fuel cycle facilities including direct disposal of spent fuel. In recognition of this potential, the Commission directed that the Advisory Committee on Nuclear Waste and Materials (the Committee) become knowledgeable concerning developments in fuel recycle and help in defining the issues most important to the NRC concerning fuel recycle facilities. The Committee decided that the most

---

[1] For the purposes of this document, "recycle" involves (a) reprocessing of the SNF (separation of the SNF into its constituent components), (b) refabrication of fresh fuels containing plutonium, minor actinides, and possibly some fission products, (c) management of solid, liquid, and gaseous wastes, and (d) storage of spent fuel and wastes.

efficient way to meet the potential needs of the Commission was to prepare a white paper on fuel recycle and chartered a group of expert consultants to do so. The paper summarized the technical, regulatory, and legal history, status, and issues related to SNF recycle to provide input to a Committee letter to the Commission and "knowledge management" (i.e., capturing the expertise of the experts preparing and reviewing this paper) concerning the history of SNF recycle and implications for current SNF recycle programs. This report was prepared to make the contents of the white paper more widely available. It is important that the reader not only understand the purposes of this paper but also realize that the paper is not intended to address the implications of advanced reactors (e.g., fast-neutron-spectrum reactors for fissioning transuranium (TRU) elements), provide detailed recycle technology descriptions and characterization, provide details on pyroprocessing, focus on fuel fabrication and refabrication, evaluate the merits of the DOE technical or programmatic approach, or provide conclusions and recommendations.

SPENT NUCLEAR FUEL RECYCLE HISTORY AND TECHNOLOGY

What Is Reprocessed?

All operating U.S. power reactors and most power reactors in the world are LWRs which are moderated and cooled with "light" (ordinary) water. The two most common types of LWRs are pressurized water reactors and boiling-water reactors. The most basic part of LWR fuel is a uranium dioxide ceramic fuel pellet which is about 1 centimeter in diameter and 2 to 3 centimeters long. The uranium enrichment is typically 3 to 5 percent ^{235}U. At some point, the fissile content of a batch of new fuel that was inserted into the reactor core is sufficiently low and the fission product content sufficiently high so that its usefulness as a power source is exhausted. At this point, the batch is removed from the reactor and sent to the storage pool as SNF. It is this SNF that constitutes the feed material for the initial step of fuel recycle/reprocessing.

How Is SNF Currently Reprocessed?

Many processes for reprocessing SNF have been developed and several have been used on a substantial scale since World War II. However, for industrial-scale applications, the only process currently being used is the PUREX (plutonium-uranium extraction) process, a diagram of which appears in Figure S.1.

The PUREX process produces the following major waste streams:

- a liquid high-level waste that would eventually be converted to glass logs for eventual disposal in a deep geologic repository

- compacted and possibly stabilized (e.g., grouted) cladding waste and undissolved solids remaining after SNF dissolution in nitric acid, which have an uncertain disposition in the United States

- waste forms containing the volatile radionuclides, which have an uncertain disposition in the United States

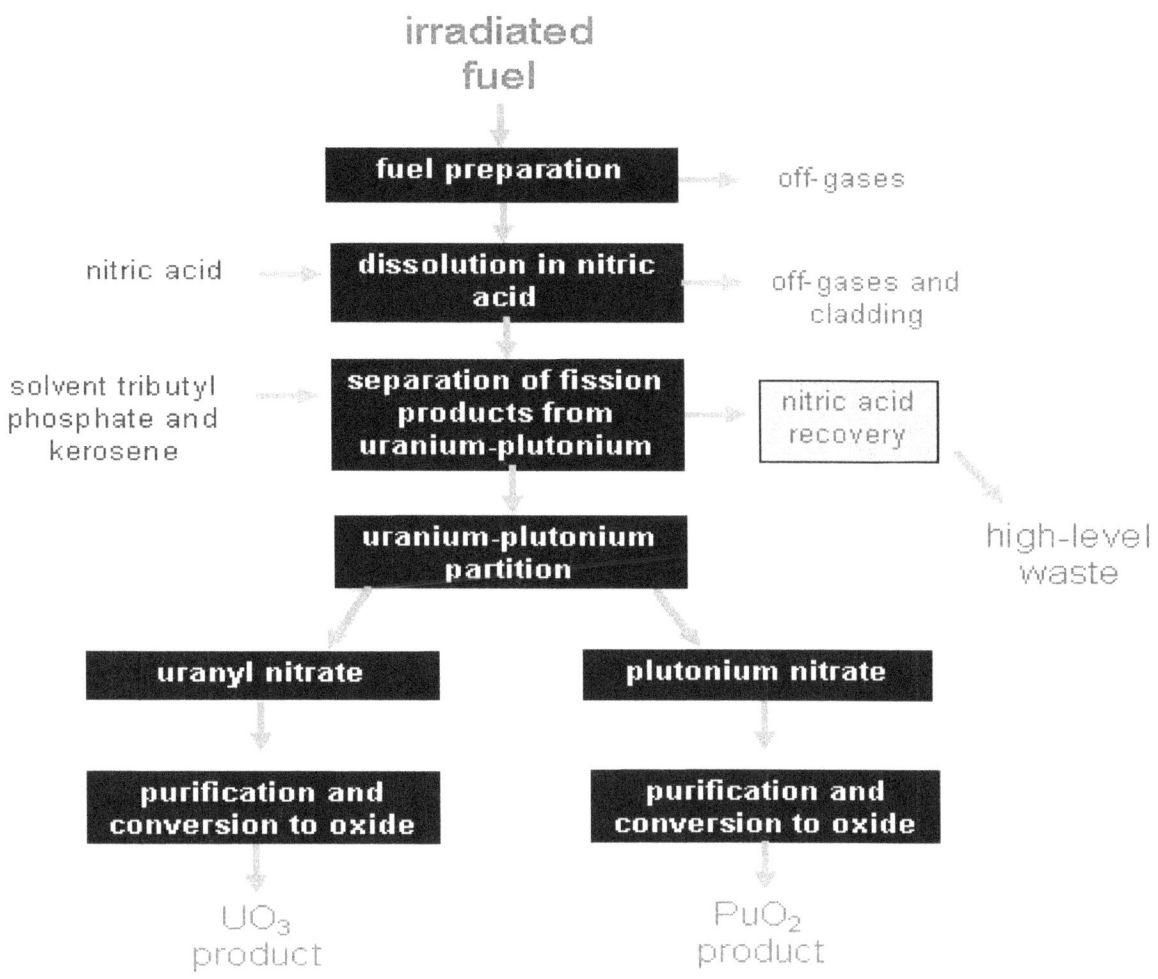

Figure S.1: Schematic diagram of the PUREX process

Where Was and Is SNF Reprocessed?

Reprocessing was carried out using the PUREX process in large Government-owned plants located in Richland, Washington, and Savannah River, South Carolina, for plutonium production. A plant was also constructed at Idaho Falls, Idaho, to recover uranium from spent naval reactor and other highly enriched fuels. These plants are no longer in operation, although some legacy nuclear materials are still being reprocessed at the Savannah River Site.

The first commercial spent fuel reprocessing plant, and the only one to operate to date in the United States, was the Nuclear Fuel Services' West Valley Plant. This plant is now shut down and undergoing decommissioning. In 1967, the U.S. Atomic Energy Commission authorized General Electric Co. to build a reprocessing plant in Morris, Illinois. However, design and operational problems caused General Electric to halt construction of the plant before it processed any spent fuel. The water pool at the site is still used to store SNF. Construction of the Barnwell Nuclear Fuel Plant in Barnwell, South Carolina, near the DOE Savannah River Site, began in 1970 but was never completed

Although the United States discontinued attempts at commercial spent fuel reprocessing in the mid-1970s, this did not deter construction and operation of reprocessing facilities worldwide. The following are the major SNF reprocessing plants in the world:

- The French La Hague spent fuel reprocessing plants UP2 and UP3 for LWR SNF have a nominal capacity of 1700 MTHM of SNF per year.

- The Thermal Oxide Reprocessing Plant (THORP) at Sellafield in the United Kingdom has a nominal capacity of 1200 MTHM of LWR and advanced gas reactor SNF per year, and the B205 plant for Magnox (metal) fuel at the same site has a capacity of 1500 MTIHM SNF per year.

- Japan has a small reprocessing plant at Tokai-mura and is beginning operation of the 800 MTHM/yr LWR SNF reprocessing plant at Rokkasho-Mura. The process used in the Rokkasho plant is largely based on French technology.

- Russia has a 400 MTHM/yr commercial reprocessing plant at Mayak.

India has three reprocessing plants, none of which is safeguarded by the International Atomic Energy Agency (IAEA). China plans to reprocess SNF and has stated [China, 1996], "China will follow Japan's lead and use the separated plutonium to fuel fast-breeder reactors."

What Is the Status of SNF Reprocessing Technology?

The many years of cumulative development and experience with SNF reprocessing in France and the United Kingdom have resulted in significant advances in simplifying the PUREX process as previously practiced and planned in this country, while achieving better and more predictable separations to the point that some of the product cleanup steps have been eliminated because they are not needed. These advances have been achieved while continuously reducing the amount of waste produced by the PUREX process to the point that the volume of waste destined for a deep geologic repository is about the same as the volume of the parent SNF fuel. This has been accomplished through careful management of facility operations, use of chemicals that can be degraded to water, nitrogen, and carbon dioxide, and the use of compactors and incinerators.

Despite the progress in optimizing the PUREX process, some approaches used in both France and the United Kingdom, although functional, may not be applicable in the United States. In particular, French and British reprocessing facilities remove volatile radionuclides from their off-gas primarily by caustic scrubbing (which captures ^3H, some of the ^{14}C, and ^{129}I) and then release these radionuclides to the sea at the end of a kilometers-long underwater pipe where they undergo massive physical and isotopic dilution.

Where Is Fuel Refabricated?

Major LWR mixed-oxide (MOX) fuel fabricators include France (MELOX, 195 MTHM/yr), the United Kingdom (Sellafield MOX Plant (SMP)), 120-MTHM/yr design capacity, 40-MTHM/yr feasible capacity), and India (100 MTHM/yr). Japan is planning a 120-MTHM/yr plant at the Rokkasho-Mura site.

An MOX fuel refabrication plant is under construction at the Savannah River Site in South Carolina to dispose of excess weapons-grade plutonium by using it for commercial power production. The NRC is licensing this facility.

ADVANCED RECYCLE TECHNOLOGY

Overview of Advanced Spent Nuclear Fuel Recycle Initiatives

The National Energy Policy [NEP, 2001] issued by President Bush in May 2001 recommended expanded use of nuclear energy in the United States, including development of advanced nuclear fuel cycles. On February 6, 2006, the Secretary of Energy launched the Global Nuclear Energy Partnership (GNEP), a comprehensive international strategy to expand the safe use of nuclear power around the world. GNEP is a broad DOE program with the goal of promoting beneficial international uses of nuclear energy through a multifaceted approach. The domestic components of GNEP are designed to address the challenges outlined in the Introduction of this Summary.

The Russians have a proposal similar to GNEP called the Global Nuclear Power Infrastructure, which calls for establishing international nuclear centers and hosting the first such center in Russia. The proposed centers would provide participating nations with full "nuclear fuel cycle services," including enriching uranium, fabricating fresh uranium fuel, and storing and reprocessing SNF [IAEA, (2007c)][2].

Advanced Fuel Reprocessing Technology

DOE proposes using a reprocessing flowsheet called UREX (uranium extraction) and has stated that it currently favors a variant called UREX+1a, although interest in UREX+2 and UREX+3 has been increasing recently. Figure S.2 shows a simplified UREX+1a flowsheet.

---

[2] IAEA (2007c). International Atomic Energy Agency, "Communication received from the resident representative of the Russian Federation to the IAEA on the establishment, structure and operation of the International Uranium Enrichment Center," INFCIRC/708. June 8, 2007.

Planning, experimentation, and evaluation of the UREX+1a process are in the early stage of development (as of early 2007). Some experiments with irradiated fuel have been carried out, but there have been no lab-scale demonstrations of the entire process using SNF or large-scale testing of key equipment using nonradioactive or uranium solutions. Such a demonstration is underway as this report is being written. Additionally, the difficulties associated with combining and operating continuously and in sequence the four distinctly different solvent extraction separations steps summarized above at one facility have not yet been addressed. Such a facility would require extensive and expensive operator training, a very complex plant, and diverse equipment types.

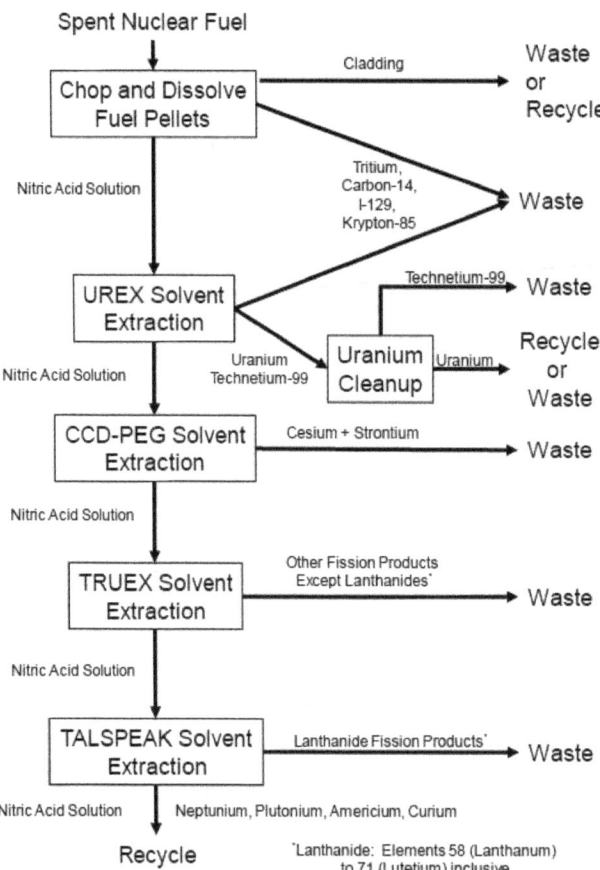

Figure S.2: Highly simplified UREX+1a flowsheet

In addition to the major wastes produced by the PUREX flowsheet (see earlier discussion), the UREX+1a flowsheet yields the following wastes:

- ^{99}Tc recovered from the uranium product stream, which is planned to be combined with the cladding waste and dissolver solids. This mixture will be compacted or melted to form an ingot. The disposition of this waste is uncertain.

- A cesium/strontium mixture that is to be made into an aluminosilicate waste form and stored in an engineered surface facility for the time required for it to decay to Class C levels (about 300 years), at which time the storage facility would be closed as a disposal facility with the cesium/strontium remaining in place.

Some consideration is being given to building a high-temperature gas-cooled reactor in the United States. Fuels for this type of reactor are distinctly different from other reactor fuels. In particular, the fuel is made mostly of graphite and is in one of two geometric configurations, either a spherical (pebble) form or a prismatic form. Reprocessing of such fuels would be similar to reprocessing LWR fuels with one important difference—a substantial quantity of graphite must be removed by burning or crushing and sieving before the fuel matrix is dissolved in nitric acid.

In the current DOE plan, pyroprocessing would be adapted to reprocessing the actinide product from UREX+1a after it had been refabricated into metallic or perhaps nitride fuel and irradiated in a transmutation reactor. Pyroprocessing, which involves the use of molten salts, molten metals, and electrochemical cells to separate SNF into its constituent parts, is inherently a batch process. After repeated batch processes, the molten salt used in the process accumulates impurities and must be discarded.

Advanced Fuel Fabrication and Refabrication

Current preparation of conventional pelletized reactor fuels for LWRs and fast reactors requires grinding the pellets to achieve a specified size and shape. This process produces finely divided fuel particles that must be recovered and recycled. A "dust-free" sol-gel microsphere pelletization process has been developed for fabrication of $(U,Pu)O_2$, $(U,Pu)C$, and $(U,Pu)N$ fuel pellets containing around 15 percent plutonium.

REGULATION AND LICENSING OF FUEL RECYCLE FACILITIES

Under current regulations, various parts of a recycle facility would have to meet the requirements of a number of regulations. The reprocessing facility per se would be licensed under Title 10, Part 50, "Domestic Licensing of Production and Utilization Facilities," of the *Code of Federal Regulations* (10 CFR Part 50). Refabrication, plutonium conversion, and recovered uranium, TRU, and cesium/strontium material storage facilities would be licensed under 10 CFR Part 70, "Domestic Licensing of Special Nuclear Material", and also under 10 CFR Part 30, "Rules of General Applicability to Domestic Licensing of Byproduct Material," (for the cesium/strontium). The uranium conversion facility would be licensed under 10 CFR Part 40, "Domestic Licensing of Source Material." The requirements of 10 CFR Part 73, "Physical Protection of Plants and Materials," and 10 CFR Part 74, "Material Control and Accounting of Special Nuclear Material," apply to all facilities.

The primary licensing regulation (10 CFR Part 50) has evolved to focus on licensing LWRs. Modifications of or exemptions from many of its requirements would be needed to accommodate the technical differences between licensing LWRs and recycle facilities.

In 2007, the Commission directed the NRC staff to begin developing the regulatory framework to license SNF recycle facilities using an option based on 10 CFR Part 70 by preparing the following:

- technical basis documentation to support rulemaking for 10 CFR Part 70 with revisions to 10 CFR Part 50 as appropriate to eliminate its applicability to licensing an SNF reprocessing plant

- a gap analysis for all NRC regulations (10 CFR Chapter I) to identify changes in regulatory requirements that would be necessary to license a reprocessing facility

The NRC has used 10 CFR Part 70 to license fuel fabrication facilities, and this regulation is currently the basis for reviewing the license application for the MOX fuel fabrication plant.

ISSUES ASSOCIATED WITH LICENSING AND REGULATING FUEL RECYCLE FACILITIES

A number of licensing or regulatory issues warrant consideration before receipt of a license application. The following sections identify these issues and provide insight into ways to address them.

Development of Licensing Regulation(s) for Recycle Facilities

Implementation of SNF recycle could involve the review of license applications for facilities that are novel in the context of the current once-through fuel cycle, including facilities for reprocessing fuels from LWRs and later for other advanced reactors, refabrication of fuels to recycle transuranium (TRU) or fission product elements or for some new reactor designs (e.g., graphite-moderated reactors), disposal of new types of wastes such as cladding and TRU (greater than Class C) waste, and extended interim storage of intermediate-lived radionuclides (cesium/strontium) followed by in situ disposal.

Modifications to important aspects of 10 CFR Part 70 would have to be considered for this regulation to be efficient and effective for licensing SNF recycle facilities. These aspects include the following:

- Use of an integrated safety analysis (ISA): 10 CFR Part 70 calls for the use of an ISA to evaluate the in-plant hazards and their interrelationship in a facility processing nuclear materials. The Committee and the Advisory Committee on Reactor Safeguards have previously recommended that a regulation based on probabilistic risk assessment (PRA) is preferable to one based on ISA because the latter has significant limitations in its treatment of dependent failures, human reliability, treatment of uncertainties, and aggregation of event sequences.

- Best estimate versus conservative approach: A companion issue to that of ISA versus PRA approaches is whether analyses should be based on data and models that represent the best estimate of what might really occur with an associated uncertainty analysis to explore the effects of incorrect data or models, or should be based on demonstrably conservative data and models. The Committee has letters on record pointing out problems with using the latter approach. Some of the most important problems arise because very conservative assumptions can mask risk-significant items, and most conservative analyses are not accompanied by a robust uncertainty analysis.

- One-step construction and operating license: 10 CFR Part 70 allows for a one-step licensing process, which means that the design and process details necessary to review the license application for a recycle facility would not be available until relatively late in the licensing process. SNF recycle facilities potentially involve equipment, chemicals, and processes that are unfamiliar to NRC staff and could lead to multiple requests for additional information from licensees and/or extensive prelicensing interactions between NRC staff and the licensee to identify and resolve potential licensing issues.

- Accommodating the potential future diversity of 10 CFR Part 70 license applications: The NRC uses 10 CFR Part 70 to license many nuclear material processing facilities other than those for fuel recycle. Such facilities are typically much smaller, less costly, and less complex than the anticipated SNF recycle facilities to the point that imposing requirements appropriate for recycle facilities could unduly burden some applicants.

- Risk-informed, performance-based[3]: In a risk-informed regulatory approach, risk provides an important insight for licensing a facility, but other considerations such as cost and environmental impacts are balanced against the required extent of risk reduction. Risk-informed regulations and licensing approaches for a wide range of situations and the opportunities for focusing scarce resources on the most risk-significant items in very complex facilities would indicate the appropriateness of a risk-informed approach in this instance. It is also prudent for regulations for licensing fuel recycle facilities to include provisions that allow the regulator to make exceptions on a case-by-case basis.

  A corollary to a regulation being risk-informed is its being performance-based. That is, the criteria for granting a license are expressed in terms of the requirements the applicant must meet but not the means by which the applicant meets the requirement. For example, a regulation that requires that a dose limit be met is performance based, but one that requires use of a specific technology is not.

- Programmatic specificity of changes to 10 CFR Part 70: Discussions concerning regulation of recycle facilities have focused on the DOE GNEP and the facilities currently being proposed by DOE. The scope, functional requirements, size, and timing of these facilities are still evolving and likely to change in unknowable ways which suggests that a more generic focus might be in order.

---

[3] In SRM-SECY-98-144, "White Paper on Risk-Informed and Performance-Based Regulation," the Commission defined risk-informed regulation in its white paper "Risk-Informed and Performance-Based Regulation" as "…a philosophy whereby risk insights are considered together with other factors to establish requirements that better focus licensee and regulatory attention on design and operational issues commensurate with their importance to public health and safety."

Impacts of SNF Recycle on Related Regulations

In addition to the need to establish the approach(es) to be used for the primary licensing regulations for fuel recycle facilities, it will be necessary to address issues that SNF recycle might raise concerning other regulations, such as the following:

- Classification of the wastes is an important determinant of their treatment, storage, transport, and disposal. Specific issues regarding waste classification include those listed below:

    - Whether the cesium/strontium wastes will require a waste determination and DOE decision considering them "wastes incidental to reprocessing" so that they would not require disposal in a deep geologic repository and criteria for reviewing a waste determination for this material.

    - The stable end point of cesium decay is stable isotopes of barium, which means that the cesium/strontium waste may be a mixed waste.

    - Uranium, ^{85}Kr, and ^{135}Cs could become wastes destined for near-surface disposal, but the waste classification tables in 10 CFR Part 61, "Licensing Requirements for Land Disposal of Radioactive Waste," do not list them.

- Determination of what constitutes an acceptable waste form and disposal technology for wastes such as cladding waste, cesium/strontium, miscellaneous wastes containing greater than 100 nCi/g TRU (e.g., equipment and analytical wastes, protective equipment, high-efficiency particulate air filters), and wastes containing ^{99}Tc, ^{129}I, and ^{14}C is necessary to define how the waste must be treated. Waste form and disposal requirements also have a significant impact on the selection of recovery processes for some species, such as those in gaseous effluents where technology selection remains open and release limits remain to be developed.

- Use of any of the UREX flowsheets for recycle would change the characteristics (e.g., volumes, forms, decay heat, penetrating radiation, and radionuclide concentrations) of the wastes going to the repository. Consequently, aspects of existing regulations and guidance concerning repository licensing that are driven by the waste characteristics (e.g., dominant contributors to repository risk, degradation rates of the spent fuel cladding and matrix, effects of penetrating radiation and decay heat on repository chemistry and water flow) may change substantially and new risk-significant licensing issues are likely to arise.

- The concentration of additional radionuclides present in recovered uranium as compared to unirradiated uranium in certain portions of enrichment equipment and wastes and the penetrating radiation from ^{232}U in the recovered uranium will have to be considered when licensing facilities for handling recycled uranium.

- Managing cesium/strontium waste by 300-year storage followed by closure of the facility as a disposal site raises the following questions:

    - Should the cesium/strontium waste be classified when it is produced or after the monitored interim storage period?

- Can a near-surface facility containing radionuclides emitting considerable amounts of heat and penetrating radiation be reliably designed, built, and maintained for as long as 300 years?

- Would such a long-term storage facility be suitable for conversion to a permanent disposal facility at that time, and what technology should be used in such a conversion?

- Construction and operation of a fuel reprocessing plant before actinide burner reactors are available would result in the need to store significant quantities of TRU elements, which raises issues about the acceptable form and technology for storing such materials product and the best means to safeguard it.

- A fundamental feature of the DOE UREX flowsheets approach is that plutonium is never completely separated from other more radioactive radionuclides. This raises issues concerning how to factor the increased radiation and difficulty in separating the plutonium into the safeguards and security paradigms that will be used in the recycle facilities.

- An important goal in licensing SNF recycle is to include design and operating requirements to minimize problems in decommissioning the facilities at the end of their operating life. A related issue is the need to obtain sufficient lessons learned to provide a basis for decommissioning requirements to be included in regulations concerning SNF recycle facilities, and how to balance these requirements against the licensee's freedom to build a plant that efficiently and economically accomplishes its mission.

- The differences among IAEA, NRC, and DOE requirements for the permissible significant (sigma) plutonium inventory differences could be important to recycle facility operation and deserve further attention.

Other Regulatory Issues Arising from SNF Recycle

The following summarizes issues that could arise from implementation of SNF recycle that could impact NRC regulations:

- The UREX flowsheets involve at least four interconnected processes operating in series. Each of these processes is as complex as the traditional PUREX process. This raises the issue of how to overcome the difficulty and resource requirements entailed in developing the technical capability (expertise, analytical tools) to evaluate whether such a complex system can be safely operated. This evaluation involves predicting the behavior of myriad pieces of equipment and the piping connecting them under normal and accident conditions.

- Recycle facilities that are capable of meeting DOE goals will involve many processes and pieces of equipment that have never been used on a commercial scale or in licensed facilities. When licensing facilities, the NRC normally performs confirmatory research to validate key data and assumptions made by a licensee. In the case of recycle facilities, such research would require highly specialized facilities (e.g., hot cells) and equipment that is available only in a limited number of places, none of which are part of the current

NRC community. The lack of NRC infrastructure relevant to SNF recycle raises the issue of how the NRC will perform confirmatory research.

- It will be necessary to create and grade licensing examinations for fuel recycle facility operators at several levels of competence and responsibility. Finding people qualified to prepare and administer proficiency examinations will be challenging.

- Regulators must complete a number of time-consuming activities before the anticipated receipt of a license application for SNF recycle facilities, including creating the licensing regulation(s) for recycle facilities, modifying supporting regulations, preparing guidance documents underpinning the foregoing, establishing release limits for volatile radionuclides such as ^{3}H and ^{14}C, and reconsidering the waste classification and disposal technology system. Establishing release limits for volatile radionuclides could be a particularly lengthy process because of the likely need to perform engineering design, cost, and risk studies as a basis for the limits.

  DOE also needs to complete several time-consuming activities before it can submit a license application for a recycle facility having the full capabilities presently envisioned by the Department (i.e., using the UREX+1a flowsheet or similar process). These activities include completing the development and testing of a complex multi-step reprocessing flowsheet, testing equipment to implement the flowsheet, developing waste treatment processes and disposal facilities for a number of novel waste streams, completing a generic environmental impact statement for the recycle program, designing the facility, and preparing the license application and other regulatory documents.

  The time required to accomplish both the regulatory and DOE activities is likely to be at least several years, but this estimate has a substantial degree of uncertainty. However, DOE could decide to initially deploy SNF recycle facilities that do not have the full capabilities presently envisioned and then add additional modules over time to achieve the full capabilities. Such an approach is significantly less complex than implementing all the envisioned capabilities at the outset and represents only a modest extension of existing technology. Consequently, the time required to develop and submit a license application could be significantly reduced compared to that needed for a fully capable facility, but the time needed for regulatory development would not be significantly reduced.

- In the 1970s, when nuclear fuel recycle was being aggressively pursued, the U.S. Environmental Protection Agency (EPA) began to develop standards for radionuclide releases from recycle facilities and codified the results in Title 40, "Protection of Environment," Part 190, "Environmental Radiation Protection Standards for Nuclear Power Operations," of the *Code of Federal Regulations* (40 CFR Part 190). With the benefit of decades of hindsight, analysis now shows that the existing standard raises the following issues:

  - The factors by which ^{85}Kr and ^{129}I must be reduced are approximately 7-fold and 200-fold, respectively. The evaluation that resulted in these factors was based on effluent control technologies that were under development but were never completed. Thus, meeting the standard with available technologies may not be feasible.

- Background information accompanying the standard indicated that studies concerning limits on releases of ^{14}C and ^{3}H were underway. These studies remain incomplete, and thus, the standard may be incomplete.

- The cost-benefit approach used in the analyses involved calculating the collective dose by integrating very small doses over very large populations and distances and comparing them to then-common metrics such as a limit of \$1000/man-rem to determine whether additional effluent controls were justified. As Committee letters and the International Commission on Radiological Protection have observed, such a comparison is questionable.

- The scope of 40 CFR Part 190 does not include fabrication of fuels enriched with plutonium or actinides other than uranium.

In summary, the EPA standard on which effluent release limits are based may impose requirements that are infeasible in the near term, may be incomplete, and is based on analysis techniques that have become questionable over the years. This is a very fragile (if not inadequate) foundation for the NRC to develop implementing regulations and begin licensing a fuel recycle facility.

- Implementing fuel recycle will require a substantial number of staff who are knowledgeable about the technical and regulatory aspects of fuel recycle facility design and operation. The design and operation of the fuel reprocessing and recycle fuel fabrication facilities are particularly challenging because staff members trained as nuclear chemical operators and engineers are required and few exist because demand in this field has been very limited for decades. This same expertise, especially that of nuclear chemical engineers, will be in demand by organizations performing fuel recycle research and development, designing and operating recycle facilities, and regulating recycle facilities, thus further exacerbating the shortfall in supply.

- GNEP goals include having once-through and recycle facilities in the United States providing services (fuel supply, fuel take-back) as a primary component. With substantial amounts of U.S. fuel going to many other countries and being returned to the this country, a more focused relationship between the NRC and regulators in other countries might be desirable or necessary to ensure that U.S. fuels are acceptable internationally and that fuel irradiated in another country has an acceptable pedigree for its return.

- DOE regulates most of its activities under its own authority, while the NRC regulates licensees engaged in civilian and commercial nuclear activities. In the case of the projected fuel recycle facilities; there is the potential for DOE regulation of some facilities that interface with other NRC-regulated facilities (e.g., a fuel refabrication plant and associated waste management facilities such as at the mixed-oxide (MOX) fuel fabrication plant at the Savannah River Site). This could pose challenges concerning compatibility and consistency of regulatory requirements, especially as it concerns material that moves between facilities and the means by which it is moved.

- The development and design of recycle facilities provide an excellent opportunity to educate and train NRC staff for licensing subsequent facilities and to obtain insights useful in developing or modifying NRC regulations to license these facilities. Of particular note is a stepwise end-to-end demonstration of the UREX+1a flowsheet now underway at Oak Ridge National Laboratory beginning with SNF receipt and ending with refabrication of fuels containing TRU elements and use of waste materials (e.g., technetium, cesium/strontium) to develop treatment processes.

RESEARCH NEEDS

Implementation of SNF recycle in the United States as presently envisioned by DOE will require information that will presumably result from the Department's ongoing research and development program or international experience. However, to fulfill its role in developing regulations and later reviewing a license application for SNF recycle facilities, the NRC staff must be able to independently assess the safety of the facilities. Such an assessment requires sufficient understanding of key technical aspects of the processes and materials in the plant. In the course of preparing the white paper, the Committee noted the following research needs that are likely to be important to the NRC's regulatory role:

- Knowledge of the split of each chemical species in each process step in the plant (the separation factors), especially concerning tritium, iodine, technetium, neptunium, and radioactive material associated with the cladding.

- Developing a model that simulates the interconnected equipment in a facility flowsheet using the separation factors to determine the radionuclide concentrations and inventory. Such models need to accommodate complexation, colloids, internal recycle streams, and important conditions in bulk fluids (e.g., temperature, acidity, radiolysis).

- Understanding stability of organic extractants, solvents, and ion exchange materials and the safety implications of degradation products.

- Understanding and documenting the technical status and cost of effluent control technologies and developing a methodology for performing the cost-benefit analysis.

- Understanding the performance of potential waste forms for krypton, iodine, carbon, technetium, and cesium/strontium in likely storage and disposal environments.

- A better understanding of the strengths, limitations, and historical performance of long-term institutional controls and facility degradation rates in the context of reviewing a license application for 300 years of near-surface storage of cesium/strontium to provide a basis for these judgments.

ACKNOWLEDGEMENTS

The authors would like to acknowledge the valuable contribution and technical insights provided by many individuals. From the nuclear industry these include Dorothy Davidson and Alan Hanson (AREVA); Alan Dobson and Colin Boardman (EnergySolutions); Eric Loewen and James Ross (GE-Hitachi); and Felix Killar (Nuclear Energy Institute). From within the government thanks goes to those individuals that took the time to meticulously review and provide comments on earlier drafts of the white paper. They include Joseph Giitter, Stewart Magruder, Phillip Reed, and James Firth (U.S. Nuclear Regulatory Commission); Buzz Savage and Daniel Stout (U.S. Department of Energy); and Ray Clark (U.S. Environmental Protection Agency). Additionally, we thank those who have made presentations to the Advisory Committee on Nuclear Waste and Materials and graciously responded to Committee questions, including James Laidler (Argonne National Laboratory) and Kemal Pasamehmetoglu (Idaho National Laboratory). Most importantly, the authors would like to thank Yoira Diaz-Sanabria for making the publication of the NUREG possible through her dedicated work and guidance.

THIS PAGE WAS LEFT BLANK INTENTIONALLY

LIST OF ACRONYMS

ABR	advanced burner reactor
ACRS	Advisory Committee on Reactor Safeguards
ACNW	Advisory Committee on Nuclear Waste (1988-2007)
ACNW&M	Advisory Committee on Nuclear Waste and Materials (2007-2008)
ADAMS	Agencywide Documents Access and Management System
AEC	U.S. Atomic Energy Commission
AFCI	Advanced Fuel Cycle Initiative
AFCF	Advanced Fuel Cycle Facility
AGR	Advanced Gas-Cooled Reactor
AHA	acetohydroxamic acid
ALARA	as low as reasonably achievable
AGNS	Allied-General Nuclear Services
ANL	Argonne National Laboratory
ASTM	American Society for Testing and Materials
ATR	Advanced Test Reactor
AVR	Arbeitsgemeinschaft versuchsreaktor (working group test reactor)
BNFL	British Nuclear Fuels Limited
BNFP	Barnwell Nuclear Fuel Plant
BWR	boiling water reactor
CANDU	Canada Deuterium Uranium Reactor
CCD-PEG	chlorinated cobalt dicarbollide-polyethylene glycol
CEA	Commissariat a l'Energie Atomique
CFR	*Code of Federal Regulations*
CFRP	Consolidated Fuel Reprocessing Program
CFTC	consolidated fuel treatment center
Ci	curie
Ci/L	curies per liter
COEX™	co-extraction
COL	construction and operating license
DF	decontamination factor
DOE	U.S. Department of Energy
DTPA	diethylenetriaminepentaacetic acid
DUPIC	direct use of spent PWR fuel in CANDU reactors
EBR-II	Experimental Breeder Reactor II
EIS	environmental impact statement
EPA	U.S. Environmental Protection Agency
ERDA	U.S. Energy and Research Development Administration
FBR	fast breeder reactor
FBRR	fast breeder research reactor
FBTR	fast breeder test reactor
FP	fission product

FRSS	fuel receiving and storage station
FY	fiscal year
GANEX	grouped actinide extraction
GCR	gas-cooled reactor
GE	General Electric Company
GEIS	generic environmental impact statement
GEN IV	Generation IV
GESMO	Generic Environmental Statement on Mixed Oxide Fuel
GNEP	Global Nuclear Energy Partnership
GNI	Global Nuclear Infrastructure
GTCC	greater than Class C
GWd	gigawatt-day
GWe	gigawatt-electric
HA	high activity
HAAR	high activity aqueous raffinate
HAN	hydroxylamine nitrate
HDEHP	bis(2-ethylhexyl) phosphoric acid
HEPA	high-efficiency particulate air
HLGPT	high-level general process trash
HLLW	high-level liquid waste
HLW	high-level waste
HM	heavy metal
HS	hot scrub
HTGR	high-temperature gas-cooled reactor
HTTR (Japan)	high-temperature engineering test reactor
HWRR	heavy water research reactor
IAEA	International Atomic Energy Agency
ICRP	International Commission on Radiological Protection
ID	inventory difference
IET	integrated equipment test
ILLW	intermediate-level liquid waste
INFCE	International Nuclear Fuel Cycle Evaluation
INL	Idaho National Laboratory
INPRO	International Project on Innovative Nuclear Reactors and Fuel Cycles
IPD	Integrated Process Demonstration
IPyC	inner pyrocarbon layer
ISA	integrated safety analysis
ISO	International Standards Organization
JAERI	Japan Atomic Energy Research Institute
JNC	Japan Nuclear Cycle
JRTF	JAERI Reprocessing Test Facility
KARP	Kalpakkam reprocessing plant
KAERI	Korean Atomic Energy Research Institute

LLGPT	low-level general process trash
LLW	low-level waste
LMF	Lead Minicell Facility
LMFBR	liquid metal fast breeder reactor
L/MTHM	liter per metric ton of heavy metal
Ln	lanthanide
LWR	light-water reactor
M	molar
MAA	material access area
mCi/L	millicurie per liter
MDF	MOX Demonstration Facility
MeV	megaelectronvolt
MOX	mixed oxide
mrem	millirem
mR/hr	millirem/hour
MSR	molten salt reactor
MT	metric ton
MTHM	metric ton of heavy metal
MTR	materials test reactor
MTIHM	MT initial heavy metal
MTU	MT uranium
MWd	megawatt days
MWe	megawatt electric
NAS	National Academy of Sciences
NEP	National Energy Policy
NFS	Nuclear Fuel Services
NPP	nuclear power plant
NRC	Nuclear Regulatory Commission
OK	odorless kerosene
OPyC	outer pyrocarbon layer
ORNL	Oak Ridge National Laboratory
PEIS	programmatic environmental impact statement
PFBR	prototype fast breeder reactor
PFDF	plutonium fuel development facility
PFFF	plutonium fuel fabrication facility
PFPF	plutonium fuel production facility
PHWR	pressurized heavy water reactor
PP	plutonium purification
PPF	plutonium product facility
ppm	parts per million
ppmw	parts per million by weight
PRA	probabilistic risk assessment
PREFRE	Power Reactor Fuel Reprocessing Facility
PUREX	plutonium and uranium recovery by extraction
PWR	pressurized water reactor
PyC	pyrolytic carbon

Q	qualified
R&D	research and development
redox	reduction/oxidation
R/hr	Roentgen Equivalent Man (REM) per hour
RIAR	Research Institute of Atomic Reactors
ROMD	Remote Operations and Maintenance Demonstration
SiC	silicon carbide
SMP	Sellafield MOX Plant
SNF	spent nuclear fuel
SRM	staff requirements memorandum
SRS	Savannah River Site
SSNM	source and special nuclear material
STP	standard temperature and pressure
TALSPEAK	trivalent actinide-lanthanide separation by phosphorus reagent extraction from aqueous complexes
TBP	tri-n-butyl phosphate
TBq	terabecquerel
THORP	Thermal Oxide Reprocessing Plant
TMPP	THORP Miniature Pilot Plant
TRISO	tristructural-isotropic
TRU	transuranium, transuranic
TRUEX	transuranium extraction
UP	uranium purification
UKAEA	United Kingdom Atomic Energy Authority
UREX	uranium extraction
VA	vital area
WSP	waste solidification plant
WTEG	waste tank equipment gallery

1. INTRODUCTION

1.1. Background and Context

The spent nuclear fuel (SNF) from existing and future nuclear power reactors in the United States poses the following challenges:

- Obtaining adequate disposal capacity for SNF and high-level waste (HLW): The United States currently has 104 operating commercial nuclear power reactors [NEI, 2007] which produce about 2100 metric ton initial heavy metal (MTIHM) of SNF each year [Kouts, 2007]. The U.S. Department of Energy (DOE) estimates that the congressionally mandated capacity limit of 70,000 MT of heavy metal equivalent imposed on the proposed Yucca Mountain repository will be committed to accumulated spent commercial fuel and other DOE wastes by about 2010 [DOE, 2006a] leading to the need for additional disposal capacity beyond this time. Other estimates [Kessler, 2006] show that if the currently planned approach to emplacing SNF in YM is maintained, the physical capacity of the site is 2.0 to 3.5 times the 70,000-MT of heavy metal equivalent legislative limit. Thus, expansion of Yucca Mountain to its physical limits could accommodate spent fuel from an additional 33 to 83 years of operation of existing nuclear power plants but proportionately fewer if reactors undergoing license extensions, new reactors similar to those presently deployed, and new types of advanced reactors were to continue or begin producing additional SNF. The characteristic of SNF that limits how much can be placed in a unit area of the repository is its decay heat, which is dominated by ^{90}Sr and ^{137}Cs for the first several decades and by certain transuranic (TRU) actinide isotopes beyond this time, with plutonium and ^{241}Am being the dominant contributors. The volume of the SNF does not drive the amount of repository area required to dispose of SNF, although the volume of SNF does affect the number of storage and shipping casks that must be handled and transported.

- Increasing utilization of available energy resources: The SNF from commercial power reactors contains two significant sources of fissile material that could be recovered and reused. The first is the ^{235}U remaining after the fuel that initially contained up to 5 percent of this isotope has been depleted. The ^{235}U concentration in SNF is typically several tenths of a percent (about the same as natural uranium) and could be reenriched to yield some additional uranium for fuel. The second significant source of fissile material in SNF is the TRU elements created by neutron irradiation of ^{235,238}U, with plutonium being the most important because it constitutes at least 1 percent of typical SNF, and about two-thirds of the plutonium is fissionable in the thermal neutron spectrum in light-water reactors (LWRs).

- Avoiding the increased proliferation risk from a pure plutonium stream: The plutonium contained in SNF has been recovered and reused in many countries. However, the processes that have been used to recover it generate the product as a stream of pure plutonium that can be handled with little or no radiation shielding and, as a consequence, poses a proliferation risk. This proliferation risk is an undesirable aspect of existing recovery processes and has impeded the reuse of plutonium.

- Reducing disposal risks from key radionuclides: SNF contains many radionuclides that could be dissolved from failed waste canisters in a closed repository, migrate to the biosphere, and constitute a risk to the public. However, only a few radionuclides have the necessary combination of longevity and mobility to be important contributors to risk [EPRI, 2003], most notably ^{99}Tc, ^{129}I, and ^{237}Np and its decay products. The neptunium in SNF is produced by neutron irradiation of ^{235}U, as well as by the decay of ^{241}Pu and ^{241}Am in the SNF that is produced by neutron irradiation of ^{238}U.

DOE has been supporting programs to recycle SNF for a number of years. Specifically, DOE is proposing to reprocess SNF (separate it into its constituent components), with LWR fuel being the primary feedstock for the foreseeable future; reuse the recovered uranium; reuse the plutonium by making it into new reactor fuel (refabrication); destroy actinides that dominate repository risk by refabricating them into fuel or targets; irradiate the actinides in a nuclear reactor; and incorporate radionuclides that cannot be readily destroyed by irradiation into appropriate waste forms. To address proliferation concerns, DOE proposes to reprocess the SNF using new approaches that do not produce a separated plutonium stream.

The current DOE program for implementing its proposed approaches is the Global Nuclear Energy Partnership (GNEP). This program contemplates building (1) an integrated nuclear fuel recycle facility,[4] (2) an advanced reactor for irradiating neptunium, plutonium, americium, and curium, and (3) an advanced fuel cycle research facility to develop the technology needed by GNEP.

In the conference report associated with the fiscal year (FY) 2006 Energy and Water Appropriations bill [Congress, 2005], Congress directed DOE to select a site for the integrated nuclear fuel recycle facility by FY 2007 and to initiate construction of one or more such facilities by FY 2010. DOE subsequently submitted a program plan [DOE, 2006a] and a strategic plan [GNEP, 2007a] providing details of its path forward and has continued to refine these plans.

Fuel recycle has the potential to require changes in the existing regulatory framework and expertise of the U.S. Nuclear Regulatory Commission (NRC) which are now structured to license LWRs and their associated once-through fuel cycle facilities including direct disposal of spent fuel. In recognition of this potential, the Commission directed [NRC, 2006 a, b] that the Advisory Committee on Nuclear Waste [or the Advisory Committee on Nuclear Waste and Materials (ACNW&M)] become knowledgeable concerning developments in fuel recycle and help in defining the issues most important to the NRC.

In FY 2006, the Committee received initial briefings on fuel recycle by Committee consultants, NRC staff, and DOE. Based on this input, the Committee decided that the most efficient way to meet the potential needs of the Commission was to prepare a white paper on fuel recycle and chartered a group of expert consultants to do so. The white paper was sent to the Commission by the ACNW&M as an attachment to a letter dated October 11, 2007 (ADAMS Accession Number ML072840119).

---

[4] For the purposes of this document, "recycle" involves (a) separation of the constituents of spent nuclear fuel, (b) refabrication of fresh fuels containing plutonium, minor actinides, and possibly some fission products, (c) management of solid, liquid, and gaseous wastes, and (d) storage of spent fuel and wastes.

1.2. Goal and Purposes

The primary goal of this report is to summarize the technical, regulatory, and legal history, status, and issues related to SNF recycle for two purposes:

(1)   To supply the basis for a letter to the Commission providing the Committee's initial insights on important regulatory issues raised by the DOE SNF recycle initiative and recommending the means and timing for the NRC to address them.

(2)   To provide "knowledge management." Because decades have elapsed since the NRC last attempted to license fuel recycle facilities, this report aims to capture the knowledge of the experts concerning the history of SNF recycle and implications for current SNF recycle programs for use by all elements of the NRC.

This report is intended to be generic and not focused exclusively on the current U.S. program directed at implementing SNF recycle (GNEP). However, if SNF facilities regulated by the NRC are built in the United States, the facilities will of course reflect a focus on the policies, goals, and priorities of the U.S. SNF recycle program as modified in the future. Consequently, important aspects of this report necessarily reflect the goals and priorities of the current GNEP program and its technology selections, because the future is unknowable. The impact of this focus is mitigated by the ambitious scope of the current GNEP program which proposes to separate SNF into a larger array of products and wastes than those produced or currently planned in other countries (France may be an exception). If some of these separations are not performed, then specific portions of this report may be academic, but the Committee believes that the major messages will be pertinent in the future.

While it is important that the reader understand the purposes of this paper, the reader should also realize that the paper is not intended to do the following:

- Address the implications of advanced reactors: This paper does not address the implications of potential new power production and/or transmutation reactors (e.g., fast-neutron-spectrum reactors for fissioning TRU elements) or devices (e.g., accelerators for transmutation) for the NRC's regulations and infrastructure. This is the purview of the NRC's Advisory Committee on Reactor Safeguards (ACRS). The paper does briefly describe the fuels that might be used in such reactors because they are the potential feed for a reprocessing plant.

- Provide detailed recycle technology descriptions and characterization: This paper does not contain detailed descriptions of the SNF recycle science or technology or the characteristics of internal plant streams for multiple reasons:

  – Such descriptions are not needed to accomplish this paper's stated goal.

  – Reliable details concerning the science and technology underlying GNEP recycle proposals are not available because the processes are still under development.

  – Where available, detailed descriptions of technology and internal plant streams are proprietary, sensitive for security reasons, or both, which would preclude the issuance of this paper as a public document.

The paper does include detailed descriptions of historical science and technology by reference.

- Provide details on pyroprocessing: If SNF recycle is to proceed, the first and largest operation will necessarily be to reprocess LWR fuel. Aqueous processes such as those currently in use internationally or advanced versions being developed in this country and elsewhere are very likely to be used on LWR fuels because they were developed for this purpose. As a consequence, this paper focuses on aqueous processes. Pyroprocesses (using molten metals and salts and electrochemical cells to accomplish SNF separation) were conceived to reprocess metal fuels and may have application to oxides and to advanced fuels such as nitrides and carbides. This paper briefly describes such processes.

- Focus on fuel fabrication and refabrication: Fabrication of new reactor fuels from the plutonium resulting from LWR fuel reprocessing and licensing of facilities for fabricating them is established practice. Many countries use uranium/plutonium oxide fuels, and a U.S. facility is in the licensing process as this paper is written. While inclusion of a mixture of TRU actinides (neptunium, plutonium, americium, and curium) does present some additional technical challenges for fabrication (e.g., much higher emission of radiation and heat), a refabrication facility for this purpose would not raise the variety of major conceptual and practical issues that SNF reprocessing does.

- Evaluate the merits of the DOE technical or programmatic approach: As stated previously, the purposes of this paper are to support preparation of a Committee letter on regulatory issues that would be raised by SNF recycle and how the NRC should address these issues and to aid in capturing knowledge that is rapidly being lost because it has not been needed in the United States for decades. Evaluation of the DOE program is the purview of appropriate elements of the executive and legislative branches, independent review groups, and other interested stakeholders.

- Contain conclusions and recommendations: A Committee letter will provide the NRC with conclusions and recommendations regarding the implications of SNF recycle.

1.3. Scope

In attempting to meet the goal and purposes stated above, this paper addresses the following topics:

- a historical overview of fuel recycle including recycle programs, reprocessing technology and facilities, and fuel refabrication technology and facilities

- a historical overview of the siting, design, operation, and material accountability of fuel recycle facilities that describes how recycle technologies were integrated into an operating facility designed to meet then-applicable (in the late 1970s) regulations and some needed improvements that were evident even at that time

4

- an overview of current recycle activities including ongoing U.S. and international fuel recycle programs, a brief discussion of reactors and the spent fuel they would generate (which is the feedstock for recycle facilities), and discussion of the advanced fuel recycle processes being developed

- an initial scoping calculation of the nature and characteristics of wastes that might result from the UREX+1a SNF reprocessing flowsheet currently favored by GNEP

- discussion of regulation and licensing of fuel recycle facilities, including the following:

  - pre-NRC experience with licensing two such facilities in the 1970s and earlier

  - discussion of regulations that might be used to license new fuel recycle facilities including existing and potential new regulations

  - topics related to licensing such as environmental protection requirements (primarily effluent controls) and other environmental impacts

  - recent proposals by the NRC staff on how fuel recycle facilities might be licensed and Commission direction related to their licensing

- a discussion of issues relevant to licensing recycle facilities, including the licensing regulation(s) per se, potential impacts on other NRC regulations, implications for NRC expertise and infrastructure, and timing

1.4. Information Sources

In addition to the many publicly available documents reviewed to prepare this white paper, other important sources of information are as follows:

- presentation by R.G. Wymer to the 171st Advisory Committee on Nuclear Waste (ACNW) Full Committee, June 6, 2006, Subject: Commercial Spent Nuclear Fuel Reprocessing

- presentation by DOE representatives to 172nd ACNW Full Committee, July 20, 2006 Subject: Advanced Fuel Cycle Initiative (AFCI)

- L. Tavlarides' trip to Argonne National Laboratory (ANL) for discussions on AMUSE code calculations, October 6, 2006

- ACNW members' trip to Hanford to tour reprocessing-like facilities, October 17–18, 2006

- J. Flack's and L. Tavlarides' trip to Idaho National Laboratory (INL), October 24–25, 2006

- presentations by R.G. Wymer, and L. Tavlarides to 174th ACNW Full Committee, November 15, 2006, Subject: White Paper on Potential Advanced Fuel Cycles

- presentations by NRC/Office of Nuclear Material Safety and Safeguards staff to 175th ACNW Full Committee, December 13, 2006, Subject: Conceptual Licensing Process for Global Nuclear Energy Partnership (GNEP) Facilities

- presentation by Government Accountability Office representative to ACNW, April 11, 2007, Subject: Scope and Methodology of the Government Accountability Office's (GAO's) Ongoing Review of the Global Nuclear Energy Partnership (GNEP) Effort

- ACNW member A.G. Croff's attendance at a briefing by DOE on the GNEP waste management strategic plan in April 2007 at the National Academy of Sciences (NAS) Nuclear and Radiations Studies Board meeting

- ACNW member A.G. Croff's attendance at May 2007 Nuclear Waste Technical Review Board meeting to hear Jim Laidler's presentation on GNEP waste streams

- presentation by AREVA representative to 179th Committee meeting, May 16, 2007, Subject: AREVA Spent Nuclear Fuel Recycle Facilities

- presentation by Energy Solutions to 181st Committee meeting, July 19, 2007, Subject: BNFL's Reprocessing Technology

- roundtable discussion with 181st Committee meeting and internal and external stakeholders, July 19, 2007, Subject: Committee White Paper on Spent Nuclear Fuel Recycle Facilities

- presentation by GE-H to the 183rd Committee meeting, October 16, 2007, Subject: SNF Recycling Processes

2. RECYCLE FACILITY FEEDSTOCK: SPENT NUCLEAR FUEL DESIGNS

This section describes the uranium-plutonium and thorium-uranium fuel cycles with emphasis on the fuels that constitute the feedstock for SNF recycle facilities.

2.1. Overview of Generic Fuel Cycles

2.1.1. Uranium-Plutonium Fuel Cycle

The uranium-plutonium fuel cycle starts with uranium ore. Historically, the uranium has been enriched to 3 to 4.5 percent in ^{235}U, although today the trend is generally to the higher enrichments (e.g., 4.5 to 5 percent). The enriched uranium is converted to oxide and fabricated into UO_2 pellets for use in reactor fuel. A portion of the ^{238}U in the fuel is converted to plutonium by capture of neutrons. Eventually, enough plutonium is produced that it contributes substantially to the fission reaction and thus to power production in power reactors. The plutonium remaining can be separated by reprocessing the spent fuel and converted to PuO_2, which is mixed with UO_2 to produce "MOX" (mixed-oxide) fuel. The advantage of this approach is that it uses the relatively abundant ^{238}U (99.275 percent) in uranium ore to produce fissile plutonium to replace part of the much less abundant ^{235}U (0.71 percent) in the fuel.

2.1.2. Thorium-Uranium Fuel Cycle

The thorium-uranium cycle starts with thorium and enriched uranium. Neutron capture in ^{232}Th produces ^{233}U, which is fissile. In principle, when enough ^{233}U is produced, it can completely replace the enriched uranium. The molten salt breeder reactor was projected to be a net breeder using the thorium fuel cycle. The Shippingport reactor[5] was operated on the thorium-uranium fuel cycle and attained a breeding ratio of about 1.01. The thorium-uranium fuel cycle has the potential to substantially reduce the consumption of enriched uranium for a given amount of energy produced.

2.2. Fuel Designs

2.2.1. Pressurized Water Reactors (PWR)

The most basic part of pressurized water reactor (PWR) fuel is a uranium oxide ceramic fuel pellet which is about 1 centimeter in diameter and 2 to 3 centimeters long. The pellets are inserted into Zircaloy cladding tubes, and plugs are welded in the end, thus constituting a fuel element or "rod." The tubes are about 1 centimeter in diameter and about 4 meters long. The gap between the fuel pellets and the cladding is filled with helium gas to improve the conduction of heat from the fuel pellet to the cladding and minimize pellet-cladding interaction which can lead to fuel element failure. The fuel elements are then grouped into a square array called a fuel assembly (see Figure 1).

---

[5] The Shippingport (Pennsylvania) breeder reactor was developed in the 1950s by the Naval Reactors Division of the U.S. Atomic Energy Commission (AEC) under Admiral Rickover.

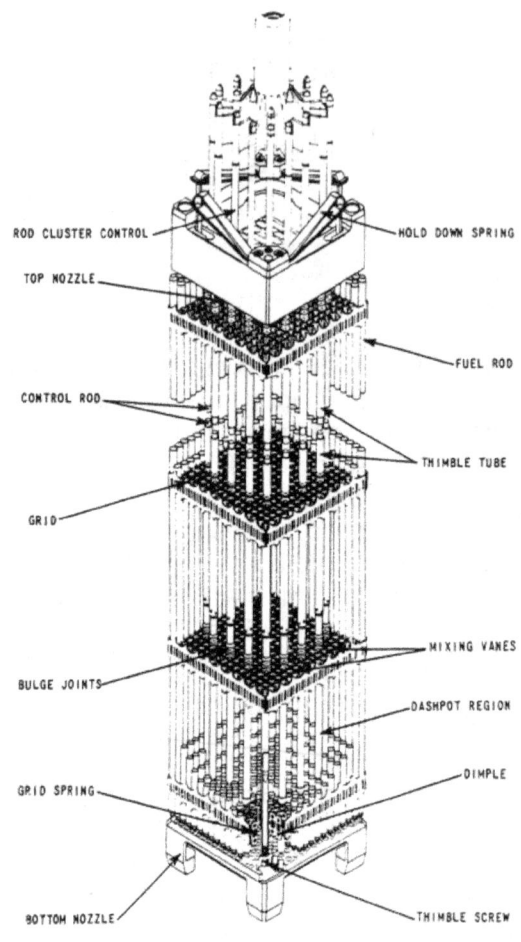

Figure 1: PWR Fuel Assembly and Hardware

There are 179 to 264 fuel elements per fuel assembly, and 121 to 193 fuel assemblies are loaded into a reactor core. The size of the fuel element array ranges from 14x14 to 17x17 rods in a square array. Typical PWR fuel assemblies are about 406 centimeters in length and 21.4 centimeters square. Control rods are inserted through the top and into the body of the assembly.

2.2.2. Boiling-Water Reactors (BWR)

In a boiling-water reactor (BWR), the fuel is similar to PWR fuel except that the assemblies are not as big in cross-section and are "canned." That is, a thin metal sheath (also known as a shroud) surrounds each assembly. The primary purpose of the sheath is to prevent local water density variations from affecting neutronics and to control the thermal hydraulics of the nuclear core. Each BWR fuel element is filled with helium to a pressure of about 3 atmospheres (300 kilopascals). A modern BWR fuel assembly comprises 74 to 100 fuel elements rods that are slightly larger in diameter than those in a PWR. There are up to 800 assemblies in a reactor core, holding up to approximately 140 MT of uranium. The number of fuel assemblies in a specific reactor is based on considerations of desired reactor power output, reactor core size, and reactor power density. Figure 2 shows modern BWR fuel assemblies and a control rod "module." The fuel element array is typically 6x6 elements to 8x8 elements. The assemblies are 10 to 15 centimeters across and about 4 meters long.

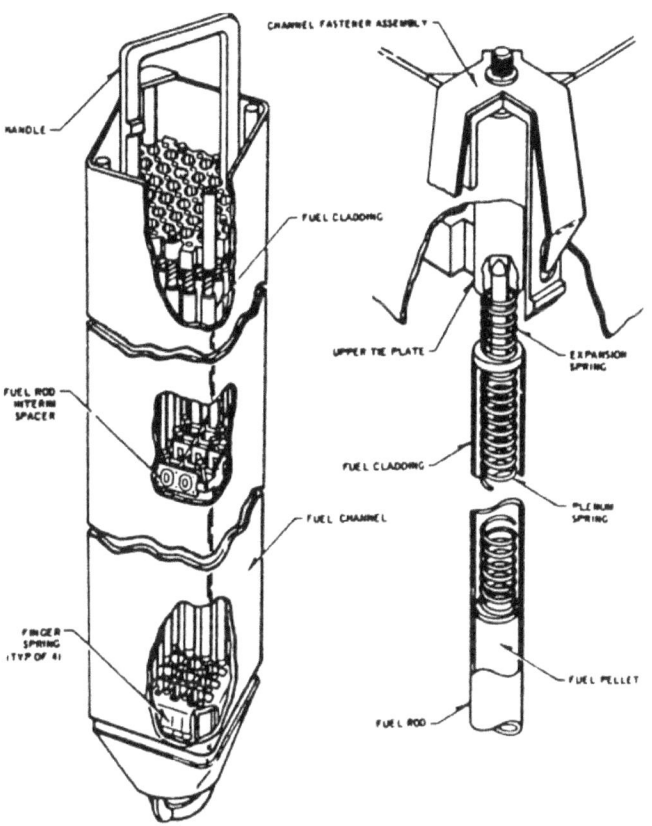

Figure 2: BWR Fuel Assembly

2.2.3. Fast Reactors

Historically, the core of a fast reactor consisted of an array of canned fuel assemblies containing an array of fuel elements. The fuel element cladding and can are both made of stainless steel which allows these reactors to operate at higher temperatures than LWRs. When such reactors were designed to produce more plutonium than they consumed (i.e., to "breed"), the core was composed of a central region of MOX fuel (called driver fuel) that could sustain a chain reaction. Above and below the driver fuel pellets were pellets of depleted uranium called a "blanket." Additionally, surrounding the driver assemblies in the radial direction were fuel assemblies in which the fuel pellets were all depleted uranium. When these assemblies are placed together, the result is creation of a central cylindrical "driver" region surrounded on all sides by the blanket. The purpose of this configuration was to use neutrons that leaked from the driver fuel to produce plutonium in the blanket.

The fuel elements are kept apart by grid spacers or in some cases by wire wound helically along each element. Driver fuel elements are typically stainless steel tubes 6 or 7 millimeters in diameter. In early designs, the elements in the blanket were larger in diameter, about 1.5 centimeters, because they require less cooling than the driver fuel elements. Both driver fuel and blanket elements may be more tightly packed in liquid-metal- (e.g., sodium, Na/K, lead, bismuth) cooled fast reactors than in LWRs because the heat transfer properties of the liquid metal are much better than those of water. This may not be the case for gas-cooled fast reactors.

In the GNEP concept, the objective of future fast reactors is to fission as many of the TRU elements as practical while still producing electricity. Thus, instead of producing about 10 percent more plutonium than what was inserted into the reactor as would have been the case with breeder reactors, DOE is seeking to have advanced burner reactors (ABRs) consume a net 25 to 75 percent of the TRU elements inserted into the reactor in the fresh fuel. Consequently, it is unlikely that there will be any blanket fuel in the ABR, and it is possible that another diluent element (e.g., zirconium) that does not produce plutonium may replace some or all of the ^{238}U in the driver fuel.

Fast reactor fuel may be made of several different materials. The principal materials are discussed below.

2.2.3.1. Oxide

Oxide fuel is made up of pellets composed of a mixture of oxides of plutonium and uranium. In the ABR, other TRU elements may be included. The equivalent enrichments[6] of the fuel range between 15 and 35 percent depending on the reactor in question. Use of oxide fuels in fast reactors is established technology.

2.2.3.2. Carbide

Historically and up to the present time, metallic and oxide fuels have been used in fast reactors.[7] There is, however, interest in the use of fuel composed of uranium/plutonium carbide, particularly in India. Carbide fuels have a higher thermal conductivity than oxide fuels and, where plutonium breeding is of interest can attain breeding ratios larger than those of oxide fuels. The increase in breeding ratio results from the fact that, while there are two atoms of oxygen per atom of uranium in the oxide, there is only one atom of carbon per uranium atom in the carbide. Light atoms such as carbon and oxygen tend to slow fission neutrons, and since there are fewer atoms per fissile atom in the carbide than in the oxide, it follows that the energy distribution of neutrons in a carbide-fueled fast reactor is shifted to higher energies than in a comparable oxide-fueled fast reactor. In addition, the density of uranium is higher in carbide fuels. The higher energy neutron spectrum and uranium density enhance plutonium production.

---

[6] Uranium and plutonium isotopes are both fissionable, so it is convenient to refer to the fissile content of fuel in terms of "equivalent enrichment" (i.e., with fissile characteristics as though it were all enriched uranium).

[7] An important exception is the fast reactor development program in India, which is based on carbide fuels.

2.2.3.3. Uranium/Plutonium/Zirconium Metal Alloy

A metal alloy of uranium/plutonium/zirconium (uranium 71 percent; plutonium 19 percent; zirconium 10 percent) in stainless steel cladding has shown considerable promise as a fast reactor fuel. It has been irradiated to burnups well over 15 atom percent [Pahl, 1990] with no deleterious effects that preclude serious consideration of its use, although some swelling and cladding interactions have been observed at these very high burnups.

2.2.3.4. Nitride

There has been interest in using uranium and/or plutonium nitride in fast reactors for many of the same reasons that carbide is attractive as a fuel. DOE is developing such fuels. An important disadvantage of nitride fuels is that they can form significant amounts of ^{14}C by neutron capture in the ^{14}N isotope of the nitrogen component. To overcome this problem, it would be necessary to perform a nitrogen isotope separation to remove the bulk of the ^{14}N

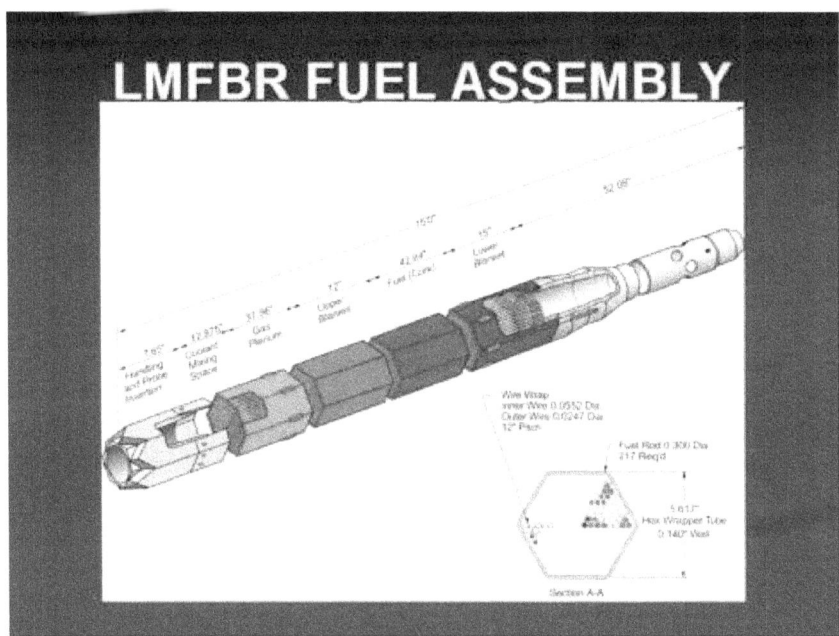

Figure 3: Drawing of a typical historical fast breeder reactor fuel assembly.
Fuel designs for the ABR are still evolving.

2.2.4. High-Temperature Gas-Cooled Reactors

The two types of high-temperature gas-cooled reactor (HTGR) fuel assemblies are spherical (called pebbles) and prismatic blocks. The former were developed in Germany in connection with the AVR and the first German HTGR power plant, the Thorium High-Temperature Reactor 300.[8] Currently, pebble bed fuel assemblies are being used in the experimental reactors HTR-10 in China and in Russia. The high-temperature engineering test reactor (HTTR) in Japan is based on prismatic fuel forms. In the United States, General Atomics developed prismatic fuels, which were used commercially in the 330-megawatt electric (MWe) Fort St. Vrain reactor.

In both cases, the fuel matrix is composed of compounds of uranium and thorium or plutonium in the form of a ceramic (usually oxide, oxycarbide, or carbide). The fuel "element" in both cases is a "TRISO" (tristructural-isotropic) fuel microsphere which is typically about 1 millimeter in diameter. TRISO fuel typically consists of a fuel kernel containing the fuel matrix in the center, coated with four layers of material. The four layers are a porous graphite buffer layer whose porosity provides space for fission gases, followed by a dense inner layer of pyrolytic carbon (PyC), followed by a ceramic layer of silicon carbide (SiC) to retain fission products at elevated temperatures and to give the TRISO microsphere more structural integrity, followed by a dense outer layer of PyC. TRISO fuel microspheres are designed not to crack because of the stresses from processes such as differential thermal expansion or fission gas pressure at temperatures above 1600 °C and therefore can contain the fuel and fission products in the worst-accident scenarios in a properly designed reactor. (See Section 3.2.3 for a detailed discussion of HTGR fuel fabrication.) These fuel microspheres are enclosed in graphite "pebbles" or prismatic graphite blocks that act as the primary neutron moderator.

2.2.5. Molten Salt Reactor

The molten salt reactor (MSR) is a unique reactor concept. It does not use a solid fuel. Instead, it uses a molten fluoride salt fuel that circulates in a loop. The loop contains a heat exchanger to extract fission energy and a system that removes fission products, primarily lanthanides and noble gases, whose presence would "poison" the salt (i.e., would capture neutrons) and ultimately prevent fission from occurring. The fuel for the Molten Salt Reactor Experiment was $LiF-BeF_2-ZrF_4-UF_4$ (65-30-5-0.1). A graphite core moderated the neutrons. The secondary coolant was $F-Li-Be$ ($2LiF-BeF_2$). At a peak temperature of 650 °C, the reactor operated for the equivalent of about 1.5 years of full-power operation.

The culmination of the Oak Ridge National Laboratory (ORNL) research during the 1970–76 timeframe resulted in an MSR design that would use $LiF-BeF_2-ThF_4-UF_4$ (72-16-12-0.4) as fuel. It was to be moderated by graphite with a 4-year replacement schedule, to use $NaF-NaBF_4$ as the secondary coolant, and to have a peak operating temperature of 705 °C. However, to date, no commercial MSRs have been built.

---

[8] South Africa has a modular pebble bed reactor under active development.

3. OVERVIEW OF SPENT NUCLEAR FUEL RECYCLE

3.1. Reprocessing Experience and Evaluations

Much of the technical information needed for reprocessing SNF and for fuel recycle in general has been available for many years and may be found in the publicly available literature. The publication dates for the general and some of the specific references at the end of this paper are indicative of the amount of detailed information available and the very long time it has been available. Notwithstanding this wealth of information, there is another component of knowledge that can only be gained through operating experience. The following sections present information based on operating experience, as it relates to early fuel recycle evaluations and the current or formerly operating recycle facilities.

3.1.1. U.S. Defense and Commercial Reprocessing Plants

In the years following World War II, Government facilities operated by DOE (formerly the AEC) carried out spent fuel reprocessing to recover plutonium for use in weapons and highly enriched uranium from naval reactor fuel.

3.1.1.1. Reprocessing for Weapons Plutonium Recovery

Large-scale reprocessing of irradiated nuclear reactor fuel to recover plutonium for use in nuclear weapons began in the United States following World War II and continued until the 1980s. Large Government-owned plants located in Richland, Washington, and Savannah River, South Carolina, carried out the reprocessing for plutonium production. A plant was also constructed at Idaho Falls, Idaho, to recover uranium from spent naval reactor and some other highly enriched SNF. The earliest large-scale plutonium recovery process was the bismuth phosphate process which was a multistep precipitation process developed by G. Seaborg and coworkers in very small-scale laboratory experiments and carried directly into large-scale production at the Hanford site in Richland, Washington. It was soon replaced with a succession of solvent extraction processes that were much simpler to operate and more efficient. These processes and the subsequent approaches used to manage them (e.g., neutralization of acidic wastes) did, however, produce copious amounts of waste, both liquid and solid, and radioactive and nonradioactive. Millions of gallons of liquid HLW were stored in large "single-shell" and "double-shell"[9] tanks on the Hanford and Savannah River sites. Most of this waste still resides in the tanks as sludge and caked salt, although efforts are underway to remove, treat, and dispose of it.

3.1.1.1.1. Bismuth Phosphate Process

The bismuth phosphate process for extracting plutonium from irradiated uranium was demonstrated in a pilot plant built next to the Oak Ridge X-10 Reactor in 1944 and subsequently deployed at Hanford. At production scale, the process produced a large amount of highly radioactive waste that contained all of the uranium in the SNF, and the bismuth phosphate process was soon replaced by a solvent extraction process. (See the following section.) The bismuth phosphate process was designed to recover plutonium from aluminum-clad uranium

---

[9] The terms "single-shell" and "double-shell" refer to whether the tanks had only one wall and bottom or whether they were, in effect, a tank within a tank. Many of the single-shell tanks have developed leaks to the subsoil.

metal fuel. The aluminum fuel cladding was removed by dissolving it in a hot solution of sodium hydroxide. After de-cladding, the uranium metal was dissolved in nitric acid. The plutonium at this point was in the +4 oxidation state. It was then carried from solution by a precipitate of bismuth phosphate formed by the addition of bismuth nitrate and phosphoric acid. The supernatant liquid (containing many of the fission products) was separated from the precipitate that contained the plutonium, which was then re-dissolved in nitric acid. An oxidant such as potassium permanganate was added to convert the plutonium to soluble PuO_2^{2+} (PuVI). A dichromate salt was added to maintain the plutonium in the +6 oxidation state. The bismuth phosphate was then re-precipitated, leaving the plutonium in solution. Then an iron salt such as ferrous sulfamate[10] was added and the plutonium re-precipitated again using a bismuth phosphate carrier precipitate as before. Lanthanum and fluoride salts were then added to create a lanthanum fluoride precipitate which acted as a carrier for the plutonium. Repeated precipitations and dissolutions were used to remove as many impurities as practical from the plutonium. The precipitate was converted to oxide by the addition of a chemical base and subsequent calcination. The lanthanum-plutonium oxide was then collected, and plutonium was reacted with nitric acid to produce a purified plutonium nitrate solution.[11]

3.1.1.1.2. Redox Process (Hexone)

The Redox solvent extraction process was used in defense SNF reprocessing facilities of the 1960s and 1970s. In this process, an acidic aqueous solution containing the dissolved SNF was contacted with an essentially immiscible organic solvent (methyl isobutyl ketone or Hexone) that preferentially removed uranium and plutonium (and, if desired, other actinides) from the aqueous phase. Many of the solvents initially employed in solvent extraction processes had significant drawbacks, such as high flammability, susceptibility to chemical and radiation damage, volatility, excessive solubility in water, high viscosity, and high cost. Solvents used in early large-scale reprocessing plants included Hexone which was used at the Hanford plant in Richland, Washington, and β,β'-dibutoxydiethylether (Butex) which was used by the British. Smaller scale applications have used bis-(2 ethylhexyl) phosphoric acid (HDEHP).

The Redox process was developed at Hanford in the late 1940s to replace the bismuth phosphate process and was used in the site's Redox plant (also known as the S Plant) from 1951 through June 1967. S Plant processed over 19,000 MTIHM of irradiated fuel. Hexone has the disadvantages of requiring the use of a salting reagent (aluminum nitrate) to increase the nitrate concentration in the aqueous phase and thus promote plutonium extraction into the Hexone phase, and of employing a volatile, flammable extractant. The aluminum in the salting agent substantially increased the volume of HLW. The Hexone, besides presenting a hazard, is degraded by concentrated nitric acid, leading to more waste as well as decreasing extraction efficiency. The Redox process was replaced by the plutonium and uranium recovery by extraction (PUREX) process.

---

[10] Ferrous sulfamate was chosen because the ferrous ion reduced the plutonium to in extractable Pu(III), and the sulfamate ion reacted to destroy any nitrous acid present. Nitrous acid had a deleterious effect on the uranium-plutonium separation process.

[11] It should be noted that large amounts of nonvolatile salts were added in the bismuth phosphate process, resulting in a large salt residue in the waste. In modern solvent extraction plants, great care is taken to eliminate as many nonvolatile salts as possible.

3.1.1.1.3. PUREX Process

These early solvents were soon replaced by tri-n-butyl phosphate (TBP), a commercially available solvent without many of the drawbacks of the other solvents. In practice, TBP is diluted about two-to-one (about 30 percent TBP) with long-chain hydrocarbons (e.g., purified kerosene or dodecane) to produce a solution with properties optimized for selectively extracting actinides. The aqueous phase in the extraction process typically is a nitric acid solution containing uranium, plutonium, neptunium, americium, curium, and fission products, most notably, cesium, strontium, iodine, technetium, and the rare earth elements (lanthanides). The plutonium and uranium (and, if desired, some other actinides by suitable valence adjustments) extract selectively into the TBP phase as complex chemical species containing nitrate ions and TBP. Adjustments of the acidity of the solution and of the valence of plutonium (from Pu(IV) to Pu(III)) make possible its subsequent separation (in a process called "stripping") from uranium. Adjustment of the valence of neptunium controls its extraction.

Adoption of the PUREX process for the production of plutonium at the Hanford and Savannah River plants for the U.S. weapons program was a major advance in irradiated fuel reprocessing. It proved to be so successful that it was adopted commercially and is the only large-scale process now used for SNF reprocessing. It has many years of demonstrated excellent performance. However, the PUREX process produces a pure plutonium stream. This may be considered a major drawback because of the nuclear weapons proliferation potential presented by separated and purified plutonium. This drawback is a major impetus for the development and adoption of new processes such as the proposed U.S. uranium extraction (UREX) processes and the French grouped actinide extraction (GANEX) process discussed in Section 6 below.

In the past, another disadvantage of the PUREX process was that it produced a relatively large amount of radioactive waste because it used plutonium-reducing agents containing nonvolatile salts such as iron compounds and because the TBP extractant contains the nonvolatile phosphate ion that leads to significant increases in waste volume. This disadvantage was not considered of much importance for weapons production but has attracted a great deal of attention in recent years in commercial plants.

In modern plants, degradable reagents are used for plutonium reduction. Steam stripping is used to remove entrained TBP and the kerosene diluent from aqueous product streams which minimizes TBP losses to waste, reduces degradation of TBP, and avoids the need for purifying the solvent by using other nonvolatile chemicals such as sodium hydroxide. It also helps prevent the conditions required for the potentially explosive "red oil" production (see Section 6.4.4). Figure 4 shows a highly simplified flowsheet for the PUREX process.

Sections 3.1.3 and 3.1.4, respectively, present additional details of the PUREX process as carried out in the Thermal Oxide Reprocessing Plant (THORP) in the United Kingdom and the La Hague plant in France.

Appendix A describes in detail the PUREX process that was to be used in the Barnwell Nuclear Fuel Plant (BNFP). Because many advances have been made in the PUREX process since the time of the BNFP, the discussion is presented primarily for historical reasons.

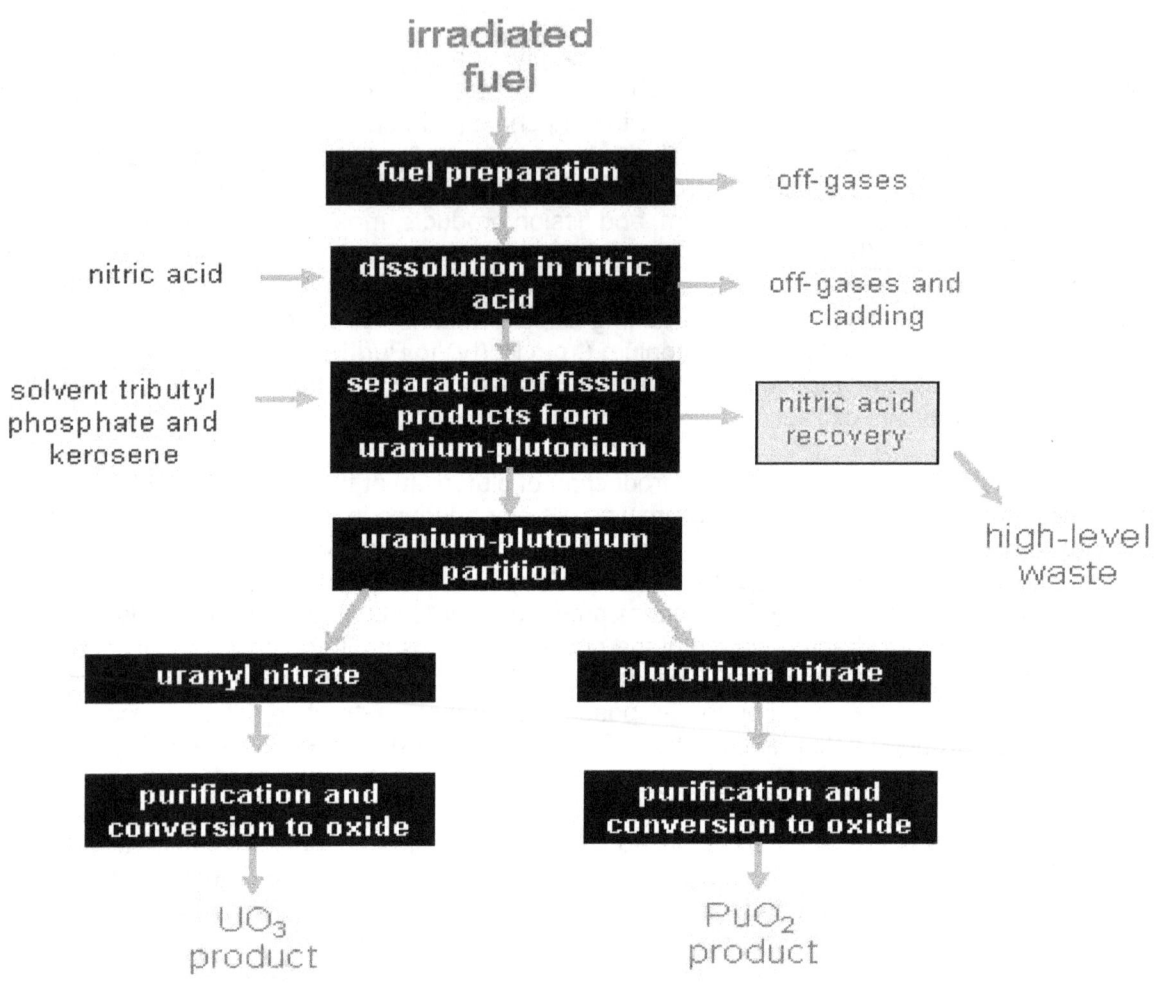

Figure 4: PUREX process flowsheet

3.1.1.2.   U.S. Commercial Reprocessing Plants

In the early days of enthusiasm for nuclear energy in general, and SNF recycle in particular, the U.S. Government encouraged commercial spent fuel recycle both in this country and overseas. As a consequence, three fuel reprocessing initiatives occurred in the United States. These are discussed briefly below.

3.1.1.2.1.   Nuclear Fuel Services West Valley Plant—operated and being decommissioned

The Nuclear Fuel Services (NFS) West Valley reprocessing plant was a 300 MTIHM per year plant that operated in western New York from 1966 until 1972 [West Valley, 1981]. Using the PUREX process, the West Valley Plant reprocessed about 650 MTIHM, about 390 MTHM of which was metallic fuel from the Hanford plutonium production reactors. Consequently, the fuel had a very low burnup of around 2000 MWd/MTHM (to be contrasted with burnups of 45,000 MWd/MTHM for today's LWR fuel). The remainder of the fuel reprocessed at the West Valley Plant was uranium oxide fuel and fuel containing thorium. Because of seismic concerns and other issues that would have greatly increased the cost, a planned expansion of the capacity of the West Valley plant was abandoned, and the plant was closed.

3.1.1.2.2.   GE Morris Plant— completed; never operated

In 1967, the AEC authorized General Electric Co. (GE) to build a reprocessing plant in Morris, Illinois. It was to employ a novel reprocessing method based on the volatility of uranium hexafluoride to separate uranium from fission products and actinides. Design and operational problems during process testing caused GE to halt construction of the plant before it processed any spent fuel. However, the plant was radioactive as a result of the performance of tests using uranium. The plant's spent fuel storage pond is currently used as an independent spent fuel storage installation to store commercial spent nuclear reactor fuel.

3.1.1.2.3.   Barnwell Nuclear Fuel Plant—nearly completed; never operated

Construction of the BNFP in Barnwell, South Carolina, near the DOE Savannah River Site (SRS), began in 1970. The projected plant capacity was 1500 MTIHM/yr. Appendix A discusses the plant design, which incorporated redundant cross-piping to accommodate possible piping failures and was based on the PUREX process. In 1976, President Ford announced that "…reprocessing and recycling plutonium should not proceed unless there is a sound reason…." [Ford, 1976] Presidents Carter's veto in 1978 of S.1811, the Energy Research and Development Administration (ERDA) Authorization Act of 1978, and his decision to defer indefinitely commercial spent fuel reprocessing effectively ended any chance for commercial operation of the plant, and it was abandoned before being licensed or operating with spent fuel (thus avoiding costly decommissioning).

3.1.2.   International Reprocessing Plant Summary

Although the United States discontinued attempts at commercial spent fuel reprocessing in the mid-1970s, this did not deter construction and operation of reprocessing facilities worldwide.

Table 1 [ISIS, 2007] summarizes the capacity of civil (non weapons) reprocessing plants that are operating or planned.

Table 1: Reprocessing Plants Operating and Planned in Other Nations

Country	Location	Scale	Rated Capacity, MTHM/yr	Feed Material
China	Lanzhou*	Pilot Plant	0.1	PWR, HWRR
France	1. La Hague UP2-800	Commercial	850	LWR
France	2. La Hague UP3	Commercial	850	LWR
India	1. Kalpakkam Reprocessing Plant (KARP)	Demonstration	100	PHWR
India	2. Lead Minnicell Facility (LMF)	Pilot Plant	n/a	FBTR
India	3. Power Reactor Fuel Reprocessing Plant (PREFRE)	Demonstration	100	PHWR, LWR
India	4. Fast Reactor Fuel Reprocessing Plant*	Commercial	n/a	FBTR
Japan	1. Rokkasho Reprocessing Plant	Commercial	800	LWR
Japan	2. JNC Tokai Reprocessing Plant	Demonstration	210	LWR
Russia	1. Research Institute of Atomic Reactors (RIAR)	Pilot Plant	1	n/a
Russia	2. RT-1, Combined Mayak	Commercial	400	VVER-440
U.K.	1. BNFL B205	Commercial	1500	U Metal (Magnox)
U.K.	2. BNFL THORP	Commercial	1200	LWR, AGR Oxide

* Undergoing commissioning.

Table 2 [ISIS, 2007] lists civil reprocessing plants that have operated in the past and have been or are being decommissioned. The relatively large number of pilot plants built before proceeding to large-scale reprocessing plants indicates the desirability of such facilities to test integrated flowsheets before plant construction and to optimize large-scale plant operations. Both France and the United Kingdom built pilot plants based on work with small-scale tests using fully irradiated fuel. Larger scale demonstration work was almost exclusively related to chemical engineering development with little or no radioactivity present other than possibly uranium.

Table 2: Decommissioned Reprocessing Plants

Country	Plant	Scale	Design Capacity, MTIHM/yr	Feed Material
France	1. Experimental Reprocessing Facility	Pilot Plant	5	
France	2. La Hague—AT1	Pilot Plant	0.365	
France	3. Laboratory RM2	Laboratory	0	
France	4. Marcoule—UP1	Defense/ Commercial	600	GCR fuels
France	5. La Hague—UP2-400	Commercial	400	GCR and LWR
Germany	Weiederaufarbeitungsanlage (WAK)	Pilot Plant	35	MOX, LWR
Italy	Eurex Pu Nitrate Line	Pilot Plant	0.1	$Pu(NO_3)_4$
Japan	JAERI Reprocessing Test Facility (JRTF)	Laboratory	-	
U.K.	BNFL B204 Reprocessing Plant	Defense	-	
U.K.	BNFL B207 Uranium Purification plant	Defense	-	
U.K.	BNFL THORP Miniature Pilot Plant (TMPP)	Pilot Plant	-	
U.K.	UKAEA Reprocessing Plant, MTR	Defense	0.02	MTR
U.K.	UKAEA Reprocessing Plant, MOX*	Defense		

* Standby plants are in decommissioned status unless otherwise noted. Not all decommissioned facilities are listed (e.g., Eurochemic in Belgium and U.S. commercial facilities discontinued in the 1970s (NFS, GE Morris, BNFP) are omitted).

3.1.2.1.   France

France has the largest LWR SNF reprocessing enterprise in the world. Commercial reprocessing is carried out at La Hague on the English Channel. La Hague reprocesses SNF from reactors belonging to French, European, and Asian electricity companies. AREVA NC La Hague (formerly COGEMA) has two operating reprocessing plants at this site (UP2-800 and UP3), each with a design throughput of 850 MTIHM of spent fuel per year. Uranium dioxide, MOX, and research and test reactor fuels can be reprocessed at La Hague. For more than 10 years, La Hague reprocessing was split between the requirements of the French nuclear program (France has 58 nuclear power plants, generating 76 percent of the country's electricity) and those of the 29 European and Japanese power companies that have reprocessing agreements with AREVA NC. Power companies from seven countries have sent or are sending spent fuel to AREVA NC La Hague (France, Japan, Germany, Belgium, Switzerland, Italy, and the Netherlands). From 1990 to 2005, close to 20,000 MTIHM of fuel were reprocessed at the La Hague site.

The UP1 reprocessing plant at Marcoule, commissioned in 1958, reprocessed 18,600 MTIHM of spent fuel from gas-cooled reactors (GCRs) and research reactors to recover the reusable nuclear materials (uranium and plutonium). The site, located in southern France close to the Rhone river, reprocessed spent fuels for Commissariat a l'Ènergie Atomique (CEA) needs (G1, G2, and G3 reactors and Chinon 1). France's commercial activities were initiated on the site in 1976, when UP1 began reprocessing spent fuel from the French natural uranium-fueled reactors, which were graphite-moderated GCRs. COGEMA was created the same year and took over the operation of the UP1 plant. Production in the UP1 plant was terminated at the end of 1997 after 40 years of operation. Since 1998, the plant has been undergoing final shutdown operations, to be followed by retrieval and repackaging of waste, then by dismantling and decommissioning of the plant.

3.1.2.2.   Great Britain

Great Britain is the second largest reprocessor of power reactor spent fuel in the world. Reprocessing is carried out at the Windscale/Sellafield plant in the northwest of England on the Irish Sea. Civilian reprocessing, which began at Windscale in 1964, is expected to continue until at least 2015, about 5 years after the shutdown of the last Magnox reactor in Britain. Magnox power reactor fuel has been reprocessed at Windscale/Sellafield since 1964. Oxide fuel reprocessing began in 1969. Large-scale oxide fuel reprocessing began with the commissioning of THORP in 1994. THORP has a nominal capacity of 1200 MTIHM of fuel per year. About 70 percent of the first 10 years of reprocessing at THORP was dedicated to foreign fuel. The British utility, British Energy, holds contracts to reprocess about 2600 MTIHM of fuel, while German utilities signed additional contracts for 700 MTIHM of fuel in 1990. In early 2005, THORP had processed almost 6000 MTIHM of SNF containing about 1.7 billion curies of radioactivity. Figure 5 shows a diagram of the current main THORP chemical separation processes. Energy Solutions provided detailed information on THORP process chemistry for inclusion in this report (see Section 3.1.3).

Fast reactor and materials test reactor (MTR) fuel has been reprocessed at Dounreay in northern Scotland since July 1958. This small reprocessing facility is now shut down and is undergoing decommissioning.

3.1.2.3.   Japan

Japan has a small reprocessing plant at Tokai-mura, with a design capacity of about 270 MTIHM per year (0.7 MTIHM/day). (The actual annual reprocessing rate has been about 100 MTIHM/yr.) Construction of Japan's first commercial reprocessing plant has been completed at Rokkasho-mura and testing for commercial startup is underway. The plant, which is primarily of French design, includes a number of buildings for the head-end process, separation and purification, uranium and plutonium co-denitration, high-level radioactive waste vitrification, and other processes related to spent fuel recycle. The plant includes many French process improvements to the PUREX process. The nominal reprocessing capacity of the plant is 800 MTIHM of uranium per year, enough to reprocess the spent fuel produced by about thirty 1000-MWe nuclear power stations.

3.1.2.4.   Russia

The primary Russian reprocessing activity is at Mayak. The Mayak nuclear fuel reprocessing plant is between the towns of Kasli and Kyshtym (also transliterated *Kishtym* or *Kishtim*), located 150 kilometers northwest of Chelyabinsk in Siberia. The plant is part of the Chelyabinsk Oblast.

In 1948, reprocessing of irradiated fuel from the Russian plutonium production reactors began at the Mayak plant. The plant underwent several modernizations and continued operation until the early 1960s. Reprocessing of irradiated fuel from the production reactors was continued at a second plant located next to the first. (The second plant subsequently was combined into a single industrial area called 235.) The second plant was adapted to extract isotopes from irradiated targets from the isotope production reactors of Chelyabinsk-65. In 1987, after two out of five production reactors were shut down, the second reprocessing plant was also shut down.

Plant RT-1 was commissioned in 1977 to reprocess spent fuel from VVER-440, BN-350, BN-600, research, and naval propulsion reactors. Most of the feed is from VVER-440 reactors. This is the only Russian facility that reprocesses spent power reactor fuel. The plant's nominal reprocessing capacity (based on spent fuel from the VVER-440 reactors) is 400 MTIHM per year. The historical average throughput of spent fuel at RT-1 is estimated to be 200 MTIHM per year. Since 1991, reprocessing of foreign spent fuel has become the main source of revenue for Mayak and has covered the cost of domestic spent fuel reprocessing. Until 1996, the Mayak Production Association had contracts with nuclear utilities from Finland, Germany, Hungary, Ukraine, and Bulgaria. By 1996, however, Bulgaria, Germany, and Finland had stopped using Mayak's services.

3.1.2.5.   India

The Indian Department of Atomic Energy operates three reprocessing plants, none of which is safeguarded by the International Atomic Energy Agency (IAEA). The plants have a total design capacity of about 200 MTIHM per year. The first Indian reprocessing plant, at the Bhabha Atomic Research Centre at Trombay, began operating in 1964 and has processed fuel from the Cirus and Dhruva research reactors. It was decommissioned in 1973 because of excessive corrosion, then refurbished, and put back into service in 1982.

A second reprocessing plant, the PREFRE facility, dedicated to reprocessing Canadian Deuterium Uranium Reactor (CANDU) Zircaloy-clad oxide power reactor fuel, was brought into operation at Tarapur in 1982. The design capacity of PREFRE is 100 MTIHM per year. However, logistical and technical problems have constrained production at the plant. Furthermore, India has sought to avoid building plutonium stockpiles. In 1995, there was a serious leak of radioactivity at the waste immobilization plant associated with the Tarapur plant.

In March 1996, cold commissioning (operation without actual spent fuel) began at the KARP located at the Indira Gandhi Centre for Atomic Research near Madras. "Hot" commissioning, with the introduction of spent fuel, was planned for the end of 1996. Originally, this site was planned to have 1000 MTIHM per year of reprocessing capacity by the year 2000, but these plans are now in limbo. The facility is currently designed to have a capacity of 100 MTIHM of CANDU fuel per year, for an annual output of about 350 kilograms of plutonium.

3.1.2.6. China

China plans to reprocess SNF, stating, "China will follow Japan's lead and use the separated plutonium to fuel fast-breeder reactors" [Kitamura, 1999]. China also plans to recycle MOX fuel for use in its PWRs and fast reactors. The China National Nuclear Corporation has announced plans to construct a facility to reprocess spent fuel with a capacity of 400–800 MTIHM per year, and China has pledged that its new plutonium extraction facilities will be open to international inspections. At present, China has a 0.1 MTIHM per year pilot plant undergoing commissioning at Lanzhou for commercial spent fuel reprocessing.

3.1.2.7. South Korea

South Korea is not expected to actually reprocess spent fuel or produce separated plutonium. However, South Korea has a collaborative program with Canada to develop the direct use of spent PWR fuel in CANDU reactors (DUPIC) process. The DUPIC program is the subject of South Korea's national case study for the IAEA International Project on Innovative Nuclear Reactors and Fuel Cycles (INPRO),[12] which is evaluating new fuel cycle technologies. The DUPIC process involves taking spent fuel from LWRs, crushing it, heating it in oxygen to oxidize the UO_2 to U_3O_8 (thus changing its crystal structure and pulverizing it) and drive off about 40 percent of the fission products (principally iodine, noble gases, tritium, cesium, and technetium), and refabricating it into oxide fuel for pressurized heavy-water reactors (PHWRs). The recycled fuel still contains all the actinides, including a plutonium content of nearly 1 percent and about 96 percent of the uranium in the initial PWR fresh fuel, which typically contains several tenths of a percent of ^{235}U. Thus, the fissile content ($^{239, 241}Pu$ plus ^{235}U) is around 1.5 percent, which is more than double that of natural uranium (0.71 percent ^{235}U), and suitable for use in today's PHWRs.

---

[12] INPRO [INPRO, 2006] is an IAEA program with the goal of providing a "Methodology for Assessment of Innovative Nuclear Energy Systems as based on a defined set of Basic Principles, User Requirements and Criteria in the areas of Economics, Sustainability and Environment, Safety, Waste Management, Proliferation Resistance and recommendations on Cross Cutting Issues." See [INPRO, 2006b, IAEA, 2003b]

3.1.3 THORP Reprocessing Plant PUREX Process

The THORP chemical separation plant was designed and constructed during the 1980s and early 1990s with a nominal capacity of 1200 MTIHM of SNF per year. The head-end facilities went into hot operation in 1994, and the chemical plant followed in January 1995. The following is an overview of the process chemistry of the chemical separation facilities in the THORP at Sellafield (see Figure 5) [Phillips, 2007; THORP, 2006; THORP, 1990a; THORP, 1984; THORP, 1992; THORP, 1990b; THORP, 1993; THORP, 1999a; THORP, 2000; THORP, 1999b]. This overview emphasizes the extraction behavior and downstream redox manipulation of consequences of ^{99}Tc and manipulation of neptunium valence which enables the effective decontamination of the uranium and plutonium products in only two solvent extraction cycles each. This reduction in the number of cycles reduces capital and operating costs and also reduces the amount of waste. The use of nonvolatile-salt-free (degradable to oxides of nitrogen, carbon, and hydrogen) redox reagents also allows nearly all wastes to be decomposed to a small volume and vitrified.

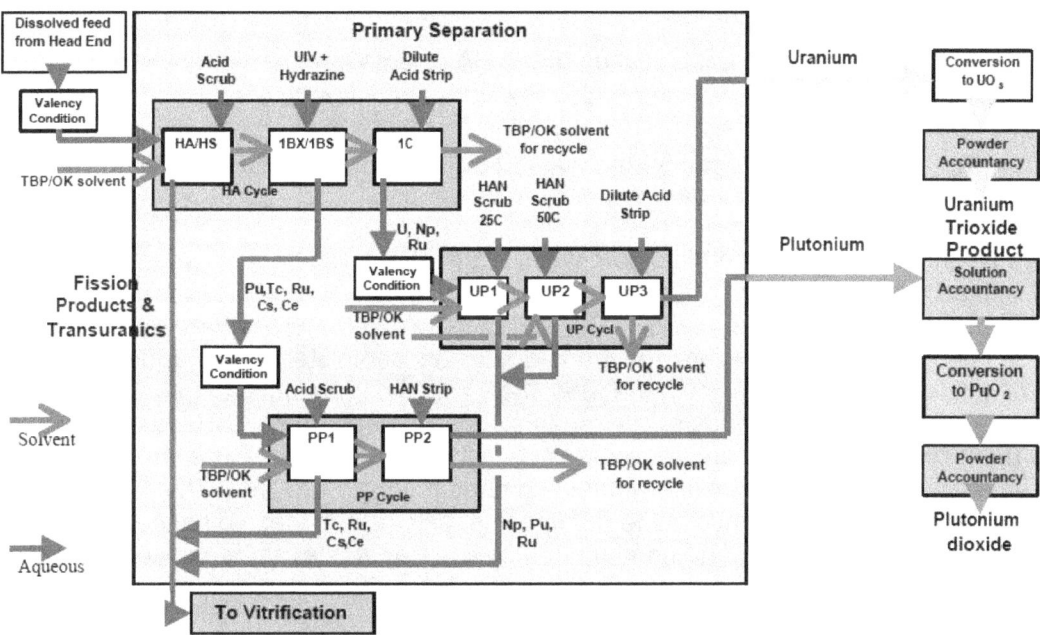

Figure 5: THORP chemical process flowsheet [Permission to use this copyrighted material is granted by Energy Solutions]

3.1.3.1.   Spent Nuclear Fuel Shearing and Dissolution

SNF is sheared into segments 1–2 inches long, and the fuel matrix is dissolved in one of three batch dissolvers in the head-end plant. The dissolver solution is clarified by the removal of undissolved fission products in one of two centrifuges and then sent on to a series of three buffer tanks of about 70 m^3 capacity each. Here the dissolver solution is adjusted to 250 grams of uranium per liter and 3-M nitric acid and treated with nitrogen oxides to ensure that all the plutonium is in the extractable [IV] valence state. Ideally, the neptunium should be in the inextractable Np(V) state so that it is not extracted (i.e., it follows the fission product waste). In practice, about 67 percent of the neptunium is in the extractable Np(VI) state.

3.1.3.2.   High Activity (HA) Cycle

The dissolver solution enters the HA pulse column at its midpoint and flows downward against an upward flowing stream of 30 percent TBP in odorless kerosene (TBP/OK). The uranium and plutonium quantitatively extract into the TBP/OK. About 67 percent of the neptunium also extracts. Almost 100 percent of the technetium extracts as a complex with zirconium.

The loaded solvent passes to the scrub section of the HA column and then onto the hot scrub (HS) pulse column, operated at 50 °C to provide maximum decontamination from ruthenium. The aqueous scrub solution is recycled to the HA column where it joins the dissolver solution and exits the bottom of the column as the HA aqueous raffinate[13] (HAAR). The raffinate is steam stripped to remove organics, evaporated, and sent to be vitrified.

The solvent containing uranium and plutonium flows to the 1BX pulse column where it is contacted with an aqueous solution of U(IV) that has been chemically stabilized with hydrazine nitrate. This reduces the plutonium to the Pu(III) state so that it transfers to the aqueous phase. Under these conditions, the technetium also transfers almost completely to the aqueous phase, but about 64 percent of the neptunium in the feed stays with the uranium in the solvent. Some uranium also transfers to the aqueous phase, so this phase passes to the 1BS pulsed column where the uranium is re-extracted and recycled back to the 1BX column. The aqueous solution of plutonium, technetium, and traces of uranium and neptunium go forward to the plutonium purification (PP) cycle. The U[IV]/hydrazine reductant is "salt free" in that it disappears after use into gaseous products (hydrazine) and uranium. This allows all wastes to be concentrated into a small volume and vitrified.

The uranium-loaded solvent, together with the bulk of the neptunium, goes to the 1C mixer-settler. This backwashes (strips), the uranium, neptunium, and traces of plutonium and ruthenium into a dilute nitric acid aqueous phase that goes to the uranium purification (UP) cycle. The stripped solvent goes to a dedicated HA cycle solvent wash system and is recycled to the HA column.

---

[13] Raffinate is the term commonly given to the portion of an input stream that remains after components have been removed in a solvent extraction separation process. In solvent extraction processes, it is the denser aqueous stream usually containing waste materials. However, some confusion may arise in the use of the term when there is an aqueous raffinate from one process step that is in fact a feed stream for a step that follows, which is the case for UREX process flowsheets.

3.1.3.3. Plutonium Purification Cycle

The aqueous feed from the 1BS pulse column is treated with oxides of nitrogen to convert the plutonium to the extractable Pu[IV] state, leaving the other components still inextractable. This stream passes to the PP1 pulse column where a fresh stream of 30-percent TBP/diluent extracts the plutonium, leaving the technetium and traces of ruthenium and neptunium in the raffinate. A scrub section at the top of the PP1 column removes impurities that are extracted along with the plutonium, with the scrub solution combining with the raffinate. This raffinate is free of nonvolatile salts and can be combined with the HAAR and sent to vitrification.

The plutonium-loaded solvent goes to the PP2 pulse column where it is contacted with an aqueous solution of hydroxylamine nitrate (HAN). HAN is an effective plutonium-reducing agent under the lower acid conditions in the PP cycle, and its use avoids the use of U(IV) (used by some other process steps for plutonium reduction), which would recontaminate the purified plutonium with uranium. The plutonium is reduced to Pu(III), transfers to the aqueous stream, and goes to plutonium dioxide production. The stripped solvent goes to a dedicated PP cycle solvent wash system and is recycled to the PP1 column.

3.1.3.4. Uranium Purification Cycle

The aqueous feed from the 1C mixer settler is conditioned at a specific temperature and acidity and for a residence time that laboratory testing showed would produce nearly 100-percent inextractable Np(V). It is then fed to the UP1 mixer-settler, where the uranium is extracted into a 20-percent TBP/diluent solvent. Neptunium stays in the UP1 aqueous raffinate. A carefully controlled HAN scrub feed is used to reduce the plutonium to Pu(III) and thus prevent its extraction, while not reducing the neptunium to the extractable Np(IV) state. The uranium-loaded solvent, with traces of plutonium and ruthenium, passes to the UP2 mixer-settler where, in the absence of neptunium, higher concentrations of HAN and higher temperatures can be used to remove the plutonium and ruthenium traces from the solvent into the UP2AR. Because this also causes some stripping of uranium, fresh solvent is fed to the uranium re-extraction section of UP2 to re-extract this uranium and combine it with the solvent from UP1. The aqueous raffinates from both UP1 and UP2 are salt-free and are routed to evaporation and vitrification along with HAAR and PP1AR.

The uranium-loaded solvent passes to the UP3 backwash (strip) contactor where dilute nitric acid is used to strip the uranium from the solvent. The stripped solvent goes to a dedicated UP cycle solvent wash process and is then recycled to UP1 and UP2.

3.1.3.5. Separation Performance of THORP

A series of conference papers have reported on the performance of THORP chemical separation (see the references above). The uranium and plutonium products have readily met international standards with the following typical overall decontamination factors (DFs):

- from the HA column feed (dissolver solution) to the uranium product
    - plutonium DF 8.6×10^6 to 1.22×10^{10}, against a flowsheet requirement of 5.0×10^6
    - neptunium DF 3.3×10^4 to 2.9×10^5, against a flowsheet requirement of 1.5×10^4
    - technetium DF 8.2×10^3 to 2.2×10^5 against a flowsheet requirement of 4.0×10^3

- for the HA column feed to the plutonium product
  - uranium DF 5.8×10^6 to 5.6×10^8 against a flowsheet requirement of 2.1×10^5
  - neptunium DF average of 6.6×10^1 against a flowsheet requirement of 4.5×10^1
  - technetium DF average of 1.0×10^2 against a flowsheet requirement of 1.0×10^2

Tables 3 and 4, respectively, show comparisons of THORP uranium and plutonium products with international specifications for recycled nuclear fuel.

Table 3: Quality of THORP UO_3 Product

Contaminant	Typical Measured Value	Specification
TRU alpha activity, Pu + Np, Bq/gU	4	≤25
Non-U gamma activity, Bq/gU	35	≤35.0*
Technetium, μg/gU	0.03	≤0.5

* Derived from American Society for Testing and Materials (ASTM) specification of less than 1.1×10^5 MeV Bq/kgU on "worst-case" basis of all activity resulting from ^{106}Ru.

Table 4: Quality of THORP PuO_2 Product

Contaminant	Typical Measures Value	U.K. Specification	ISO Specification, 1996
Uranium, μg/gPu	12	≤1000	Report
Fission products, Bq/gPu	650	≤3×10^5	Report
Nonvolatile oxides, μg/gPu	170	≤5000	≤5000

3.1.3.6. Neptunium Chemistry in THORP

Neptunium exists in nitric acid solution in three valence states—extractable Np(IV) and Np(VI) and inextractable Np(V). Typical uranium-plutonium separations using strong redox reagents to produce inextractable Pu(IV) therefore tend also to produce extractable Np(IV), which thus follows the uranium stream.

During the development of the THORP UP cycle, considerable research was done to understand neptunium redox behavior. Researchers found that a combination of careful neptunium valence control and the use of HAN-reducing agent enabled neptunium Np(V) to be maintained in the presence of Pu(IV), thus giving good decontamination of both of these species from the uranium.

3.1.3.7. Technetium Chemistry in THORP

THORP development work using actual irradiated SNF showed that 100 percent of the technetium present in the feed was extracted. This was unexpected in that previous alpha-active trials showed only about 30 percent co-extracted with the uranium. Studies found that the zirconium present in actual SNF (and not present in the alpha-active trials) complexed with the technetium to form an extractable species in the HA column and that the zirconium was then scrubbed out in the HS column and recycled to pick up more technetium. Technetium stayed in the organic phase through complexation with the uranium.

In THORP, 100 percent of the technetium was allowed to go forward to the 1B system where detailed flowsheet and equipment changes were made to cope with its effect on the hydrazine stabilizer and hence the uranium/plutonium separation efficiency. In flowsheets that require separation of the technetium as a separate waste stream, the 100 percent extraction is useful in that it provides the opportunity to include a high-acidity technetium scrub contactor immediately after the HS contactor.

3.1.3.8. Summary

THORP uses modern salt-free redox reagents together with flowsheet chemistry to produce excellent decontamination of both uranium and plutonium in three cycles of solvent extraction. This minimizes the production of waste streams. The salt-free nature of the reagents also means that nearly all waste streams can be evaporated to small volume and vitrified.

3.1.4. La Hague Reprocessing Plant PUREX Process

The following information was provided by AREVA [AREVA, 2007a; Davidson, 2007; Phillips, 2007] for use in this report.

The French La Hague reprocessing plants (UP2 and UP3) and the Japanese Rokkasho reprocessing plant, which is an evolutionary improvement over the UP3 plant, are designed to reprocess LWR spent fuel and have a design life of 50 years. Sufficient flexibility is built into the plants to accommodate spent LWR fuel with high burnups, as well as research reactor fuel and MOX fuel.

The La Hague reprocessing steps are basically the same as those in all reprocessing plants. However, there are substantial process differences among the plants based on operating experience and preference. The La Hague plant UP3 process steps are discussed below:

3.1.4.1. Spent Fuel Receiving and Storage

The AREVA La Hague plant uses two spent fuel unloading processes, underwater unloading designed for 110 casks per year and dry unloading designed for 245 casks per year. Dry unloading has the advantages of reduced worker radiation dose, quicker unloading, and a 5-fold reduction in effluents per cask unloading

The La Hague spent fuel storage capacity is approximately 14,000 MTIHM, which is about eight times the plant annual spent fuel treatment capacity.

3.1.4.2. Shearing and Dissolution

Spent fuel assemblies are cut into segments with a shearing machine that is located above a continuous dissolver. The pieces fall into a perforated basket in the dissolver where the fuel matrix dissolves in nitric acid but the cladding does not. The dissolver design is geometrically safe to avoid inadvertent criticality. When MOX fuel is dissolved, a neutron poison is added to the solution. Cladding hulls are rinsed and sent to a facility for compaction and conditioning as intermediate-level waste.[14] Any residual solids remaining in the dissolver solution are removed by centrifugation. The following improvements to the shearing and dissolution steps are being pursued:

- techniques for managing precipitates in the dissolver and development of chemical and mechanical processes to clean the dissolving equipment

- better understanding of corrosion to establish a proven and significant lifetime for the principal dissolving equipment

- adaptation of reprocessing facilities to accommodate higher burnup fuel, MOX fuel, research and test reactor fuel, and unirradiated fast breeder reactor (FBR) fuel

3.1.4.3. Uranium/Plutonium Solvent Extraction Separation and Purification

Solvent extraction with TBP in a branched dodecane diluent is used to remove uranium and plutonium from other actinides and from fission products. A nitric acid scrub is used to remove impurities carried into the TBP. Two extraction cycles in pulse columns, mixer-settlers, or centrifugal contactors are needed to meet product specifications. At the end of the extraction, scrubbing, and stripping cycles, the following solutions are produced:

- uranyl nitrate

- plutonium nitrate

- raffinates containing most nonvolatile fission products and the minor actinides

- the TBP/diluent solvent, which is treated to remove impurities and recycled

Particular attention was paid to solvent cleanup. Vacuum distillation was a major innovation that ensured purification of used solvent for recycle back into the process line.

Pulse columns were selected for use in the most highly radioactive parts of the plant, mainly to comply with criticality safety requirements. Pulse columns for solvent extraction were superior to mixer-settlers because of the shorter residence time of radioactive solutions in pulse columns. This led to drastic reductions in solvent degradation and to improved management of interfacial cruds.

---

[14] The United States does not have an intermediate-level waste classification. Such wastes would typically be greater than Class C (GTCC) low-level waste (LLW).

The UP2-800 La Hague plant[15] has three extraction cycles—one for co-decontamination and separation of uranium and plutonium, and one each for further purification of uranium and plutonium. One alkaline solvent regeneration unit is associated with the uranium and plutonium cycles. The organic solvent is recycled after cleanup by vacuum distillation. The UP3 plant was initially commissioned with two UP cycles. It subsequently became apparent that increased understanding of solvent extraction chemistry and better process control made the second UP cycle unnecessary. The second cycle ceased operation in 1994. Figure 6 shows the original and current UP3 solvent extraction cycles.

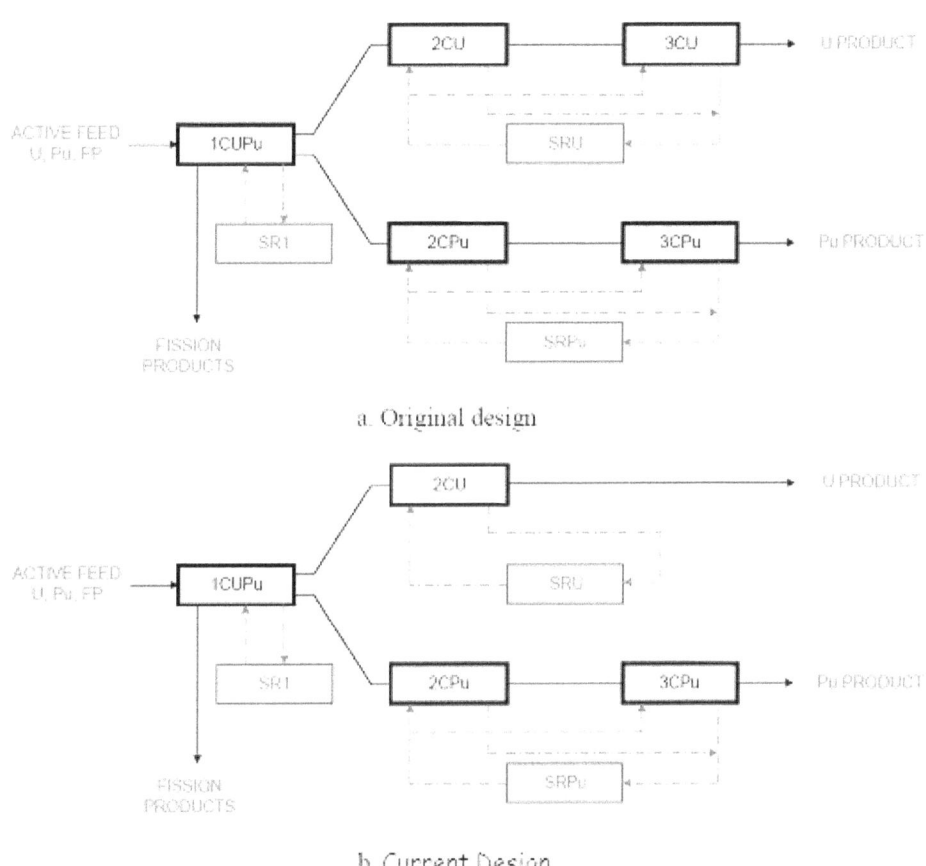

a. Original design

b. Current Design

Figure 6: Comparison of original and current French UP3 reprocessing plant solvent extraction cycles

_____

[15] The Rokkasho reprocessing plant in Japan also has three solvent extraction cycles.

3.1.4.4. Conversion of Uranium and Plutonium to Products

The uranium solution is concentrated by evaporation, stored, and eventually shipped off-site for conversion. The plutonium is precipitated as the oxalate by the addition of oxalic acid. The precipitate is filtered, dried, and calcined to form PuO_2 that meets the specifications for making MOX fuel. The mother liquor containing dissolved or suspended plutonium is concentrated and recycled.

3.1.4.5. Management and Treatment of Process Wastes

Process waste streams include the following:

- hulls and end pieces from the dissolver that are compacted for final disposal

- high-activity liquid waste streams containing the following:
    – suspended particles from feed clarification
    – fission products and minor actinides
    – concentrates generated by evaporation in the acid recovery units

The various streams, except the suspended particles, are concentrated and stored in tanks fitted with cooling and pulse devices to keep solids suspended. The concentrates are mixed with the suspended particles and vitrified to form a glass waste form.

3.1.4.6. Radioelements Released

The principal radioelements released from the plant are listed below:

- Most of the tritium is trapped in tritiated water which is released to the sea.

- About a third of the ^{14}C, which is present as CO_2, is scrubbed from the off-gas by passing it through a sodium hydroxide solution, diluted in tritiated water, and released to the sea with the remaining two-thirds being released to the atmosphere.

- Most of the iodine (^{129}I is the isotope of concern) is scrubbed by passing it through a sodium hydroxide solution which is released to the sea. Any remaining gaseous iodine is trapped in filters.

- ^{85}Kr is not removed from the off-gas stream.

- Aerosols are trapped on filters with a 99.9-percent efficiency. Ruthenium in vitrification off-gas aerosols is removed by injection of nitrogen oxides before being released. Aerosols released from the facility consist mainly of ruthenium and antimony.

Table 5 shows the La Hague UP3 plant liquid releases of important radionuclides for 2006 [AREVA 2007b]. Table 6 shows gaseous releases [AREVA 2007b].

Table 5: La Hague Reprocessing Plant Radionuclide Liquid Releases to the Sea in 2006

Radionuclide	TBq* released	TBq yearly limit in France (Ci)	% of limit
Tritium	11100	18,500 (5e+5)	59.81
^{14}C	7.46	42 (1.13e+02)	17.76
Radioiodine	1.34	2.60 (7.03+01)	51.62
^{90}Sr	0.216	2 (5.4e+01)	10.8
^{134}Cs	0.0605	2 (5.4e+01)	3.03
^{137}Cs	0.623	2 (5.4e+01)	31.15
^{106}Ru	4.8	15 (4.05e+02)	31.98
^{60}Co	0.21	1 (2.73+01)	21
Other β and γ	5.24	30 (8.10e+02)	17.45
α	0.025	0.1 (2.7e+00)	25.01

* TBq: terabecquerels (10^{12} disintegrations per second); 1 terabecquerel = ~37 curies

Table 6: La Hague Reprocessing Plant Radionuclide Gaseous Releases to the Atmosphere in 2006

Radio-nuclide	TBq released	TBq yearly limit in France (Ci)	% of limit
Tritium	67.8	150 (4.05e+03)	45.22
Radioiodine	0.00681	0.02 (5.4e-01)	34.04
Noble gases	242000	470,000 (1.27e+07)	51.58
^{14}C	14.2	28 (7.56e+02)	50.7
Other β and γ	0.000106	0.0010 (2.7e-02)	10.6
α	0.0000173	0.00001 (2.7e-04)	17.3

These tables show that all releases from La Hague reprocessing are less than the allowable release limits in France. Additionally, radionuclide releases from the La Hague plant to the atmosphere are in general much less than those from aqueous discharges from the plant. ^{14}C, which is released as CO_2 and is a soft beta emitter, and the noble gases are exceptions. The noble gases have short half-lives.

3.1.5. Accidents at Spent Fuel Reprocessing Facilities

3.1.5.1. Sellafield Facility [Schneider, 2001]

In 1973, the Windscale plant experienced a release of radioactive material following an exothermal chemical reaction in a reprocessing tank. This accident involved a release of radioactive material into a plant operating area.

In 2005, a radioactive leak from a pipe between the dissolver and a tank in the THORP fuel reprocessing plant was detected. This resulted in an extended shutdown of the facility for repairs, government investigations, fines, and potential legal charges against plant managers.

3.1.5.2. La Hague Facility [Schneider, 2001]

On October 2, 1968, ^{129}I was released through the UP2-400 stack. This accident was caused by the treatment of insufficiently cooled graphite fuels.

On January 14, 1970, the temperature of the chemical dissolution reaction of graphite fuel increased sharply, and an explosion due to hydrogen gas caused release of radionuclides including ^{129}I.

On January 2, 1980 there was a leak 200 meters from shore through a 1-meter crack in the La Hague discharge pipe that extends kilometers out to sea.
On February 13, 1990, there was an uncontrolled release of ^{137}Cs by the ELAN II B plant chimney. Routine replacement of a chimney filter led to the release of nonfiltered and contaminated air for 10 minutes.

Since 1983, corrosion of metallic waste stored in concrete pools that leaked has resulted in release of radionuclides to ground water and nearby streams; ^{90}Sr has been the most prominent of these radionuclides.

3.1.5.3. Mayak [Azizova, 2005]

In 1957, one of the concrete HLW waste storage tanks' cooling systems broke down, which permitted the tank to go dry and overheat. Chemical reaction of dry nitrate and acetate salts in the waste tank containing highly active waste caused an explosion that contaminated an area later called the "Kyshtym footprint."

On April 6, 1993, a tank containing a solution of paraffin hydrocarbon and TBP used to process spent nuclear reactor fuel exploded. The resulting explosion was strong enough to knock down walls on two floors of the facility and caused a fire.

3.1.5.4.   Tokai Reprocessing Plant [NNI, 1997]

In March 1997, a fire and an explosion occurred at the Tokai waste bitumenization facility. The accident contaminated 37 workers, and an area of 1 km^2 around the plant was evacuated.

3.1.5.5.   International Nuclear Event Scale and Accident Classification

Table 7 [IAEA, 2001] shows the existing International Nuclear Event Scale. This table indicates on a scale of 1 to 7 the severity of a nuclear accident or incident, along with a description of the nature of the event, which is currently used to categorize nuclear events. Users of this scale need to consider the relative risk from radionuclides as compared to ^{131}I to determine the category of an event. The IAEA is currently revising the International Nuclear Event Scale [IAEA, 2007b]. Table 8 [Schneider, 2001] gives specific examples of accidents that have occurred.

Table 7: The International Nuclear Event Scale

Level/ Descriptor	Nature of Event	Examples
ACCIDENTS		
7 Major accident	External release of a large fraction of the radioactive material in a large facility, in quantities radiologically equivalent to more than tens of thousands of terabecquerels[a] of ^{131}I.	Chernobyl, USSR
6 Serious accident	External release of radioactive material in quantities radiologically equivalent to the order of thousands to tens of thousands of terabecquerels of ^{131}I and likely to result in full implementation of countermeasures to limit serious health effects.	Kyshtym reprocessing plant, USSR
5 Accident with offsite risk	External release of radioactive material in quantities radiologically equivalent to the order of thousands to tens of thousands of terabecquerels of ^{131}I and likely to result in partial implementation of countermeasures to lessen the likelihood of health effects.	Windscale Pile, UK Three-Mile Island
4 Accident without significant offsite risk	External release of radioactivity resulting in a dose to the critical group of the order of a few millisieverts. Significant damage to the nuclear facility. Irradiation of one or more workers which results in an overexposure where a high probability of early death occurs.	1973 Windscale Reprocessing Plant, UK 1980 Saint-Laurent NPP France

Continuation of Table 7.

INCIDENTS		
3 Serious incident	External release of radioactivity resulting in a dose to the critical group of the order of tenths of millisieverts. Onsite events resulting in doses to workers sufficient to cause acute health effects and/or an event resulting in a severe spread of contamination (e.g., a few thousand terabecquerels), but releases in a secondary containment where the material can be returned to a satisfactory storage area. Incidents in which a further failure of safety systems could lead to accident conditions if certain initiators were to occur.	1989 Vandellos NPP, Spain, 1989
2 Incident	Incidents with significant failure in safety provisions but with sufficient defense in depth remaining to cope with additional failures. An event resulting in a dose to a worker exceeding a statutory annual dose limit and/or an event which leads to the presence of significant quantities of radioactivity in the installation in areas not expected by design and which require corrective action.	
1 Anomaly	Anomaly beyond the authorized operating regime but with significant defense-in-depth remaining.	

[a] 1 terabecquerel = 27 Ci

Table 8: <u>Types and Occurrences of Accidents at Reprocessing Plants and Sites</u>

Type of Accident	Liquid Releases	Gaseous Releases	Occurrence
Criticality in dissolver tank	X	X	Windscale, 1973 Tokai, 1999*
Fire		X	La Hague, 1981 Karlsruhe, 1985 Tokai, 1997
Explosion		X	Savannah River, 1953 Kyshtym, 1957 Oak Ridge, 1959 La Hague, 1970 Savannah River, 1975 UTP Ontario, 1980 Tomsk-7, 1993 Tokai, 1997 Hanford, 1997
Leak of a discharge pipe; breach in a tank	X		La Hague, 1979-80 Sellafield, 1983
Loss of coolant		X	Savannah River, 1965 La Hague, 1980

*The September 1999 accident at Tokai-Mura did not involve a reprocessing plant but is a type of accident which could occur in a reprocessing plant.

3.1.6. Consolidated Fuel Reprocessing Program

One of the earliest integrated attempts by the U.S. Government to develop and deploy civilian fuel recycle technology was the Consolidated Fuel Reprocessing Program (CFRP). CFRP was initiated in 1974 at ORNL primarily to advance the technology of fast reactor fuel reprocessing, although many aspects of the technology were applicable to all conventional fuel reprocessing. The program emphasis was on process automation technology, robotics, process computerization, and head-end process steps to improve gaseous effluent control.

Automation technology has been widely adopted in the manufacturing industry and in the chemical processing industries but, until recently, only to a limited extent in nuclear fuel reprocessing. It is, however, widely used in LWR fuel fabrication, especially concerning chemical conversion processes for uranium. The effective use of automation in reprocessing had been limited by the lack of diverse and reliable process instrumentation and the general unavailability of sophisticated computer software designed specifically for reprocessing plant process control.

The CFRP developed a new facility, the Integrated Equipment Test (IET) Facility, in part to demonstrate new concepts for control of nuclear fuel reprocessing plants using advanced instrumentation and a modern, microprocessor-based control system. The IET Facility consisted of the Integrated Process Demonstration (IPD) and the Remote Operations and Maintenance Demonstration (ROMD). The IPD focused on demonstration of state-of-the-art equipment and processes, improved safeguards and accountability, low-flow cell ventilation, advances in criticality safety and operability, and new concepts for control of nuclear fuel reprocessing plants using advanced instrumentation and a modern, microprocessor-based control system. The ROMD served as a test bed for fully remote operations and maintenance concepts and improved facility layout and equipment rack designs. This facility provided for testing of all chemical process features of a prototypical fuel reprocessing plant that can be demonstrated with unirradiated uranium-bearing feed materials. The goal was demonstration of the plant automation concept and development of techniques for similar applications in a full-scale plant. It was hoped that the automation work in the IET facility would be useful to others in reprocessing by helping to avoid costly mistakes caused by the underutilization or misapplication of process automation.

During the 1970s and 1980s, CFRP was a leader in advancing technology used in fuel reprocessing. The program established many contacts with foreign governments such as those of the United Kingdom, France, Germany, Japan, Russia, and Korea to share information and establish policy.

Eventually, the CFRP became reliant on the infusion of money from the Japanese nuclear enterprise and onsite Japanese technical personnel for survival. Because of the moratorium imposed by the Carter administration on U.S. reprocessing, much of the U.S.-supported CFRP technology that was developed has to date found more application in Japan than in the United States.

3.1.7. International Nuclear Fuel Cycle Evaluation

President Carter's April 1977 statement on nuclear policy that made a commitment to defer indefinitely the commercial reprocessing and recycling of plutonium—coupled with low prices for fossil fuels and uranium—effectively ended consideration of nondefense recycle activities in the United States for decades. However, the immediate result of the deferral was the initiation of a series of studies to evaluate the need for reprocessing and plutonium recycle. The largest of these was the International Nuclear Fuel Cycle Evaluation (INFCE).

INFCE addressed essentially all the important technical issues related to fuel recycle. In October 1977, the INFCE Committee was initiated, mainly at the urging of the United States, to investigate opportunities to safely internationalize the nuclear fuel cycle. INFCE participants met between 1977 and 1980 to address ways to use the nuclear fuel cycle to produce nuclear energy with a reduced risk of nuclear proliferation by modifying the fuel cycle technological base. INFCE highlighted a number of measures to counter the dangers of nuclear proliferation, including institutional and technical measures, as well as improvement and further development of IAEA safeguards. Subsequent to INFCE, reprocessing of SNF and recycle of the resulting nuclear materials was virtually ignored in this country until the turn of the century.

3.1.7.1.   Content of the International Nuclear Fuel Cycle Evaluation Study

INFCE focused on (1) an overall assessment of the nuclear fuel cycle, (2) measures to improve assurances of availability of plutonium supply for reactor fuels to developing states, (3) SNF storage, (4) improvements to nuclear safeguards, and (5) alternatives to an international nuclear economy based on plutonium and highly-enriched uranium fuels. Concern about introduction of an international plutonium economy led the IAEA in 1978 to establish a Committee on International Plutonium Storage. Establishment of this committee was the principal recommendation of INFCE.

3.1.7.2.   Principal Conclusions

The report of INFCE Working Group 4 [INFCE, 1980], one of eight INFCE working groups, discussed reprocessing, plutonium handling, and recycle of plutonium to thermal reactors. Working Group 5 addressed fast reactor recycle, and Working Group 8 dealt with other fuel recycle concepts (e.g., thorium-based and research reactor fuels). The basic recommendation of INFCE was to deposit plutonium surplus to national needs with the IAEA. This strategy for controlling plutonium envisioned that excess plutonium would be placed under international inspection and control until needed for use in civil nuclear power applications.

3.2.   Fuel Fabrication and Refabrication

Fabrication of fresh fuel and refabrication of fuel from reprocessed SNF are an international industry. A recent IAEA publication [IAEA, 2007a] gives information on both the characteristics (e.g., the ^{99}Tc concentration) and the specifications of reprocessed UO_3.

3.2.1. Fuel Refabrication Technology

The fuels for LWRs are of two types, (1) low-enriched uranium oxide and (2) mixed uranium-plutonium oxides (MOX). The uranium oxide fuels are much more common, but as more plutonium becomes available, MOX fuels are becoming more widespread. Both fuel types are made from what are essentially the dioxides of the two fissile metallic components.

For UO_2 pellet material fabrication, uranyl nitrate solution is denitrated in a fluidized bed or rotary kiln to form UO_2. Plutonium nitrate solutions are treated similarly to uranyl nitrate solutions if PuO_2 is sought. For MOX fuel material preparation, uranium and plutonium oxide powders are blended, or uranium and plutonium solutions are mixed, concentrated, and simultaneously denitrated (by microwave heating) to produce a mixed uranium/plutonium oxide (MOX). UO_{2+x}, PuO_{2+x}, and MOX are then treated by the following steps:

(1)   They are calcined in air at 800 ºC.

(2)   The calcined product is heated in a reduction furnace in H_2/N_2 at 800 ºC to produce UO_2, PuO_2, or MOX fuel material suitable for pellet fabrication. (This two-step reduction saves hydrogen.)

(3)   The powders are blended when appropriate and mixed with volatile binders.

(4)   After pressing and sintering to form pellets, the pellets are ground to meet specifications.

(5)   The LWR fuel pellets are inserted into Zircaloy cladding tubes which are grouped into a square array with grid spacers and held together with two stainless steel end pieces connected by empty tie rods. Zircaloy, an alloy of zirconium, is used for neutron economy. It has a low cross-section for capture of neutrons in the thermal neutron energy spectrum found in LWR cores.

(6)   Fast reactor fuel is fabricated using stainless steel cladding and hardware. Stainless steel is suitable for use with liquid metal coolants and high temperatures. Neutron economy is not as important in fast reactors where the neutron energy is higher than in LWRs resulting in smaller neutron absorption cross-sections.

Figure 7 [Ayer, 1988] shows a diagram of the steps in conventional MOX fuel refabrication. Other refabrication processes have been developed and deployed. Summary descriptions of these processes show that they typically differ in the details of how the uranium and plutonium oxide powders are blended [IAEA, 2003a].

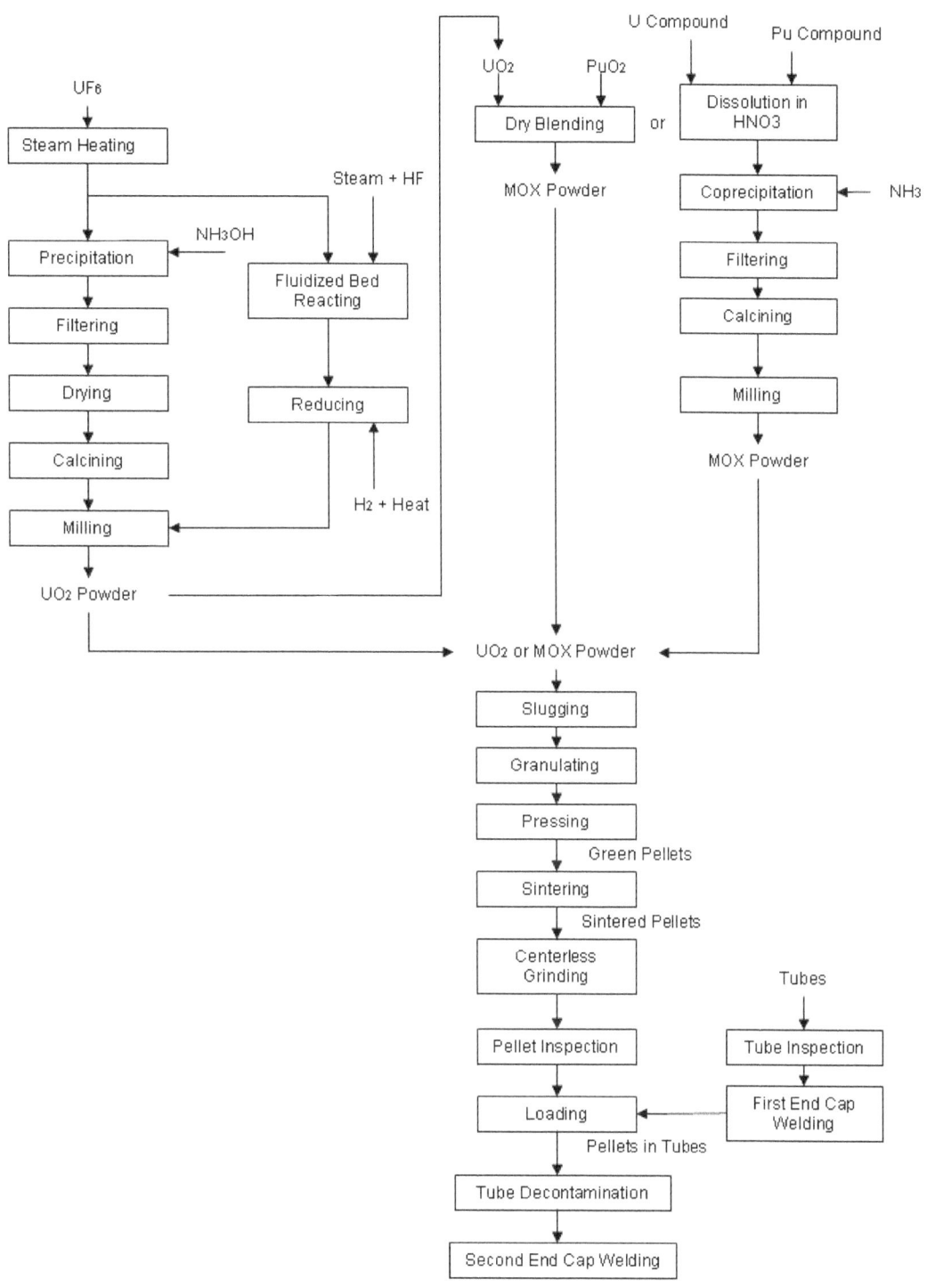

Figure 7: Diagram of MOX fuel fabrication process

3.2.2. Mixed-Oxide Fuel Fabrication Facilities

Table 9 [ISIS, 2007] lists the capacity and status of MOX fuel fabrication plants worldwide. In addition to the MOX fabrication plants listed in Table 9, DOE has an MOX plant under construction at the SRS in South Carolina. The facility is to be built as part of the national strategy to dispose of excess weapons-grade plutonium by using it for commercial power production and then disposing of the resulting SNF. The facility is to be used only for the purpose of disposition of surplus plutonium and for the plant to be subject to NRC licensing. The current plan is for the facility to be shut down when the weapons plutonium disposition is completed.

A recent IAEA document provides details of MOX fuel fabrication worldwide [IAEA, 2003a].

3.2.3. High-Temperature Gas-Cooled Reactor Fuel Fabrication

HTGR fuel is very different from other types of solid reactor fuels, and fabricating HTGR fuel is entirely different from fabricating LWR or fast reactor fuels. Both Germany and the United States have developed HTGR fuel fabrication processes for HTGR TRISO fuel particle (see Section 2.2.1) preparation that consist of a number of similar steps. In both countries, kernels containing the fissile material are made via a sol-gel process,[16] followed by washing, drying, and calcining to produce spherical UO_2 kernels (in Germany) and UCO kernels (in the United States). The major difference in the processes consists of a sintering step using CO in the U.S. process to ensure the requisite C/O stoichiometry in the kernel. The coating processes for the inner porous carbon "buffer" layer are similar, based on chemical vapor deposition from a mixture of argon and acetylene in a fluidized coater operating between 1250 and 1300 °C. A 5-micron seal coat is added in the U.S. process to seal the porous buffer coating, but this step does not occur in the German process. Table 10 gives typical properties of coated fuel particles and pebbles. Figure 8 is a schematic diagram and photograph of TRISO fuel particles.

---

[16] In sol-gel processes, a colloidal suspension (sol) is "gelled" to form a solid by extraction of water and addition of a mild chemical base. When the process is carried out using droplets of sol, spherical gel particles are formed.

Table 9: Capacity and Status of Operating MOX Fuel Fabrication Plants

Country	Plant	Scale	Design Capacity, MTHM/yr	Product Material
France	Melox	Commercial	195	MOX for LWRs
India	Advanced Fuel Fabrication Facility	Commercial	100 (nominal)	MOX for BWR, PFBR
India	Kalpakkam MOX Breeder Fuel Fabrication (under construction)	Commercial	—	MOX for PFBR
Japan	JNC Tokai (PFDF-MOX)	Laboratory	0.03	MOX fuel element
Japan	JNC Tokai (PFFF-ATR)	Pilot Plant	10	MOX fuel assembly
Japan	JNC Tokai (PFPF-FBR)	Pilot Plant	5	MOX fuel assembly
Japan	Rokkasho MOX Plant (planned)	Commercial	120	MOX for LWRs
Russia	Mayak-Paket	Pilot Plant	0.5	FBRR MOX fuel
Russia	Research Institute of Atomic Reactors	Pilot Plant	1	FBR (Vibropack)
U.K.	Sellafield MOX Plant	Pilot Plant (MDF)	Likely 40	MOX for LWRs
U.K.	Sellafield MOX Plant	Commercial (SMP)	120 design 40 feasible	MOX for LWRs

41

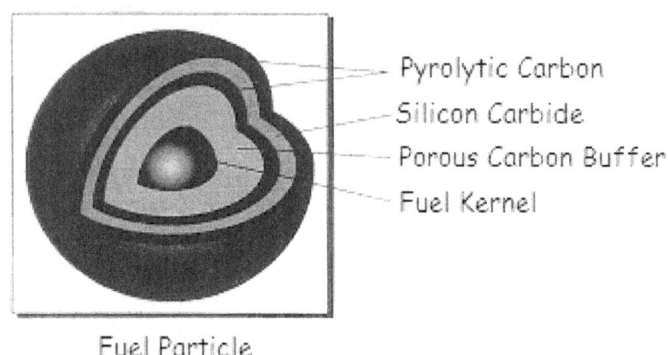

Pyrolytic Carbon
Silicon Carbide
Porous Carbon Buffer
Fuel Kernel

Fuel Particle

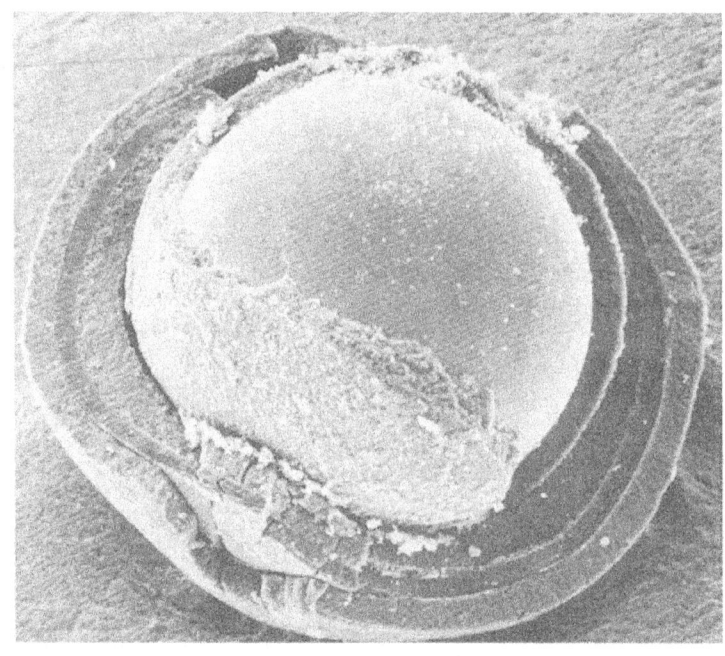

**Unirradiated TRISO Coated Particle
Broken to Show Layers**

Figure 8: Schematic diagram and photograph of TRISO fuel particles

Table 10: Typical Coated Particle Composition and Dimensions for Pebble Bed Fuel

Microspheres

Kernel composition: UO_2
Kernel diameter: 501 µm
Enrichment (^{235}U wt.%): 93
Thickness of coatings (µm):
      Buffer   92
      Inner PyC   38
      SiC   33
      Outer PyC   41
Particle diameter: 909 µm

Pebbles

Heavy metal loading (g/pebble): 6.0
^{235}U content (g/pebble): 1.00 ± 1%
Number of coated particles per pebble: 9560
Volume packing fraction (%): 6.2
Defective SiC layers (U/Utot): 7.8×10^{-6}

Figure 9 shows a diagram of a "pebble" of the type used in the pebble bed reactor.

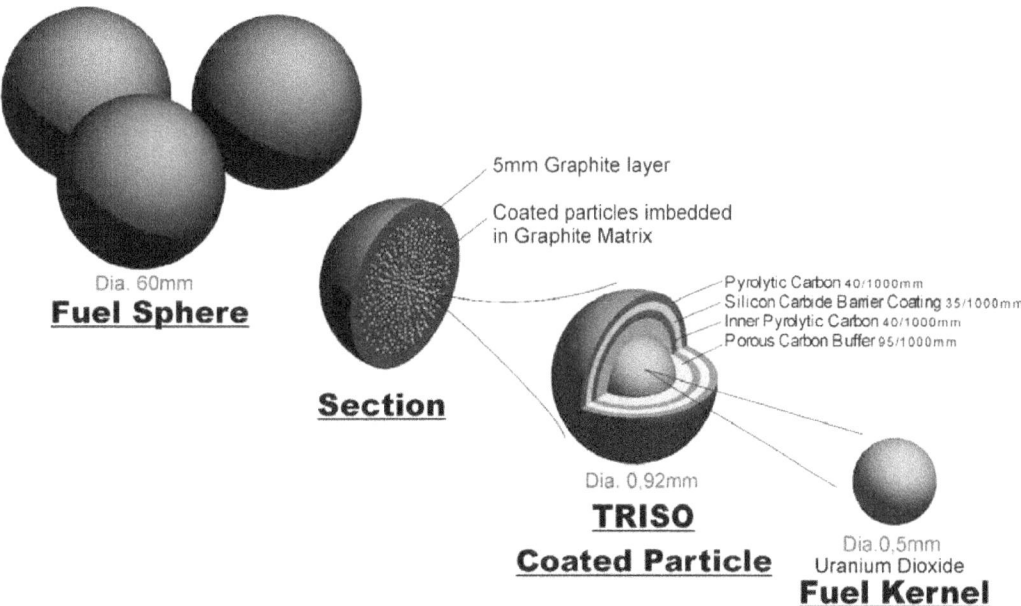

Figure 9: Pebble bed reactor fuel element

Figure 10 shows a prismatic fuel assembly of the type developed by General Atomics and used in the commercial Fort St. Vrain power reactor. These fuel assemblies are 14 inches from one flat vertical face to the opposing face and about a meter high. Fuel "sticks" of pyrolyzed carbon containing TRISO fuel particles are inserted into holes in the fuel block. Channels extend completely through the prismatic block for coolant gas flow. Larger channels provide openings into which boron carbide control rods may be inserted.

A major difference in the production of the TRISO coating is that all three layers are coated in a continuous manner in the German process, whereas in the U.S. process the fuel particles are unloaded from the coater after each coating layer to perform quality control measurements. The inner pyrocarbon (IPyC) layer in both cases is deposited from a mixture of acetylene, propylene, and argon. The temperature in the U.S. process is somewhat lower than in the German process, and the coating gas concentration is different, producing a different microstructure and density for the IPyC layer. The SiC layer is deposited from a mixture of hydrogen and methyltrichlorosilane at similar coating rates, although the temperature for the U.S. coating is about 150 °C higher than that used in the German process. The outer pyrocarbon layer (OPyC) layer is coated in a manner similar to the IPyC layer. In the United States, a seal coat and protective pyrocarbon layer are added. These layers are not counted in determining that the particle is a "TRISO" particle.

The fuel pebble in Germany uses graphite powder and organic binders to produce a powder matrix to contain the particles and to create the spherical fuel pebble. In the United States, a thick liquid matrix composed of petroleum pitch, graphite flour, and graphite shim mixed with organic binders is used to make the fuel compact. Both fuel forms are pressed and carbonized at high temperature (800–900°C).

Ultra-high purity systems and feedstock are used in the manufacture of pebbles in Germany to ensure adequate control of impurities. Both fuel forms undergo a final heat treatment, with the U.S. compact heated at 1650 °C and the German pebble at 1800 to 1950 °C in vacuum.

Figure 10: Prismatic HTGR fuel element

4. RECYCLE FACILITY SITING AND DESIGN

The primary purpose of a reprocessing plant is to chemically separate the fissile content of irradiated nuclear material from fission products and other actinide elements to recover fissile (^{235}U, ^{239,240}Pu, ^{233}U) and fertile (^{238}U, ^{232}Th) radionuclides. The five major steps in building and operating a reprocessing plant are (1) site selection, (2) plant design, (3) plant construction, (4) plant operation, and (5) waste management.

4.1. Site Selection

Many considerations determine the siting of a reprocessing plant. These include proximity to reactors providing the spent fuel, geology, hydrology, seismology, climatology, flooding potential, topography, demographics, and land uses in the surrounding area (e.g., agriculture, industry, and transportation). These considerations are discussed in more detail below.

Proximity to reactors producing the spent fuel is important from the point of view of reducing radiation exposure during transportation and increasing the security of shipments but, under some circumstances, may not be of paramount importance. This would be true, for example, if the spent fuel were of foreign origin. In any case, shipment of the spent fuel to the reprocessing plant may be cause for public concern.

Geology of the site is important if radioactive liquid effluents are released because the nature and conformation of the soil strongly influence the rate of transport of radionuclides through the environment. For example, clay has an affinity for important radioisotopes such as ^{137}Cs and rare earths and is likely to be self-healing if fractured, whereas granite has little such affinity, and there is no tendency for cracks to heal. Additionally, it is desirable to build a reprocessing plant where background radiation is low and not highly variable because it is difficult to establish an environmental monitoring radioactivity baseline where radiation levels are high or fluctuate widely. This problem can occur where uranium or thorium levels in the soil are high, leading to high radon levels that may produce large radiation background variations during climatic inversions.

Hydrology is very important if radioactive liquid effluents are released because the predominant mechanism for transport of radionuclides is movement via ground water. (In the case of an accident, transport by air can become of great importance as, for example, in the Chernobyl accident.) Aqueous transport may occur via the mechanism of water carrying dissolved ions of radionuclides or colloids (e.g., colloids of plutonium) or pseudocolloids of iron or clay to which radionuclides are sorbed. The aqueous pathway is a source of non-natural radiation dose to the public through direct ingestion of radionuclides or through contamination of agricultural products that have been irrigated using contaminated ground water obtained from wells or streams. Hydrology may also be an important consideration in supplying water for use in the facility if there are no nearby sources of plentiful surface water.

Seismology has a major impact on the licensing of plant sites and on plant construction. The plant must be sited where it is practical, both economically and physically, to ensure and demonstrate that its integrity can be retained during a projected earthquake of reasonable probability. Those parts of the reprocessing plant that contain heavy shielding and contain the highest levels of radioactivity must be capable of withstanding earthquakes with no loss of containment integrity.

Climatology plays a role in plant siting because some areas are prone to seasonal weather extremes, such as hurricanes, tornadoes, snow and ice storms, and fires in dry weather.

Flooding potential is an important consideration if the site is located in a flood plain, near rivers or streams, or is in the path of seasonal snow-melt runoff or dam failure.

Topography plays a role because the cost of plant construction may be high if grades are too steep, too much soil removal is required, or water drainage is inadequate and poses construction and subsequent operational problems.

Demographics play a major role in gaining public acceptance of a site. Whenever practicable, it is desirable to site a reprocessing plant distant from large population centers. This consideration may be at odds with the aim of locating the reprocessing plant near reactors to minimize transportation problems and is an example of often conflicting siting considerations.

Agriculture and industry in the neighborhood of a potential plant site can be of considerable importance. The presence of a facility that handles large amounts of radioactivity can be claimed to diminish the value of the crops, the land, or the agricultural and industrial products of the area. Additionally, release of radioactivity and concomitant contamination of expensive crops or industrial buildings and machinery can lead to very large financial obligations.

Transportation activities, such as commercial air, rail, or truck traffic, must be considered. This applies both to the transport of radioactive materials and to ordinary commercial traffic. Heavily traveled highways, such as interstates, in the immediate vicinity of the plant may cause concern to the public or the departments of transportation, both Federal and local. Intermodal spent fuel transport, including use of navigable waterways, may raise concerns among sportsmen, as well as health departments, if the waterways are the source of drinking water. These transportation issues are especially nettlesome because of the need to balance negative public perception with the desire for the plant to be reasonably close to the source of the spent fuel.

4.2. Design and Construction

A typical spent fuel reprocessing facility is designed and constructed to minimize the release of radioactive materials within and outside the facility both during routine operation and under unusual or accident conditions. Specifically, the current Title 10, Part 20, "Standards for Protection Against Radiation," Subpart E, "Radiological Criteria for License Termination," Section 1406, "Minimization of Contamination," of the *Code of Federal Regulations* (10 CFR 20.1406) states the following:

> Applicants for licenses, other than renewals, after August 20, 1997, shall describe in the application how facility design and procedures for operation will minimize, to the extent practicable, contamination of the facility and the environment, facilitate eventual decommissioning, and minimize, to the extent practicable, the generation of radioactive waste.

At least two physical barriers (and frequently more than two) contain the radioactive materials within the facility during operation. These barriers are typically the process equipment (vessels, pipes, etc.) and the building around the processing equipment. In most cases, the building itself provides two barriers—the hot cell or room where the process equipment is located and the outer building shell.

The following discussion is based mostly on the BNFP. International experience has contributed to significant advances in the design and operation of reprocessing plants.

4.2.1. Design

Historically, recycle plants have consisted of four major processing facilities plus a fuel receiving and storage area:

(1)   *The separations facility*, in which the spent fuel assemblies are processed to recover uranium and plutonium as nitrate solutions and where the bulk of radioactive byproduct wastes are separated as a concentrated nitrate solution of HLW.

(2)   *The uranium hexafluoride facility* in which the recovered purified uranyl nitrate solution is converted to UF_6 suitable as a feed material for isotopic re-enrichment if desired.

(3)   *The plutonium product facility* in which the recovered plutonium nitrate solution is converted to PuO_2, suitable for use in the production of MOX.

(4)   *Waste management facilities* for the handling, stabilization, packaging, assaying, inspection, and interim storage of waste before shipment to a disposal facility appropriate for each type of waste.

The NRC's Office of Nuclear Regulatory Research is developing guidance to implement 10 CFR 20.1406 to facilitate decommissioning of nuclear facilities licensed after August 20, 2007. The goal of this guidance is to ensure throughout the life of the facility that design and operating procedures minimize the amount of residual radioactivity that will require remediation at the time of decommissioning. This guidance will apply to reprocessing plants.

The actual design of these major facilities will be directly related to the regulations effective at the time of licensing and the desired/required form of both the fissile material and the waste material discharged. Proliferation and safeguards are of national and international concern when considering the construction of a recycling plant, as are attacks by terrorists. Beyond these overriding considerations, very important practical matters must be taken into account in the design, construction, and operation of a plant.

It is necessary to optimize the plant configuration for reprocessing to minimize the overall facility capital and operating costs. This is done by considering the interplay of many factors. Initial decisions include whether the plant is to be designed with a single, multiple-step process line or whether it will have parallel process lines. If the plant is to process a variety of fuel types or a very large throughput is required, then parallel lines will facilitate processing dissimilar fuel types, allow maintenance of one line when the other line is in operation, or allow practical equipment sizes while achieving high throughput.

Another aspect of plant optimization concerns approaches for waste treatment (e.g., concentration of liquid wastes by evaporation and compaction or melting of spent fuel cladding hulls and other hardware), storage, and disposal. Design optimization also addresses radiation protection of workers through use of the minimum shielding thickness consistent with meeting as low as reasonably achievable (ALARA) and radiation dose and radioactivity confinement requirements; appropriate selection of the processes carried out in the plant; and careful choice of the equipment used to carry out those processes. Simple, reliable equipment, continuous operation where possible, and ease of remote removal and replacement of equipment all contribute to minimizing capital and operating costs. In addition to being able to achieve the desired throughput, each equipment piece in the high-radiation areas of the plant must be capable of being replaced remotely or have a very low probability of failure (e.g., have no moving parts, be exceedingly corrosion resistant, be critically safe, be matched to the characteristics of the fuel assemblies to be reprocessed, and be chosen insofar as possible to be of standard sizes). The likely causes of inoperability of a reprocessing plant are the structural failure of equipment or piping in an inaccessible area as a result of corrosion or mechanical failure or failure of some part of the separation process.

There is an optimum point in the design of criticality features. For example, there is a tradeoff between having many small, critically safe process lines that offer protection for dissolver feed through geometry and having fewer, larger lines that achieve criticality safety through other means such as neutron poisons. The choice is made largely on the basis of cost, with the option of a large number of smaller lines being more costly.

Some general guidelines apply to plant design. It is desirable for radiation protection and ease of operation to put equipment for receiving the spent fuel, spent fuel pool and HLW storage, fuel segment storage, and reprocessing product storage in separate cells interconnected through transfer channels to the processing area. Ventilation and waste treatment capabilities may be provided separately for each segment. However, some facilities, such as those used in maintenance, may be shared. Avoiding inaccessible equipment or piping is also very important.

Another area that has proved troublesome is managing the complexity and cost associated with different fuel types and sizes. Variable fuel designs require different handling equipment for casks and fuel assemblies and interim storage racks or casks.

4.2.2. Construction

Process equipment should be fabricated from materials that are resistant to corrosive failure and that operate very reliably. Process equipment designed to prevent major releases of radionuclides under conditions assumed to be credible was designated as being of "Q" design.[17] These "Q" systems must provide confinement integrity for design-basis accidents and naturally occurring events such as earthquakes and tornadoes. In other less critical areas, the design membrane stress of the equipment had been established at 80 to 90 percent of the yield stress during a design-basis earthquake. Structural barriers are designed to contain process materials if primary equipment barriers are breached. The principal structural barriers are constructed of heavily reinforced concrete.

The structural barriers for process equipment, generally termed "radioactive process cells," are usually surrounded by maintenance or operating areas. The process cells where the spent fuel is chopped and dissolved and where high-level liquid wastes are concentrated have very high radiation levels. At BNFP, these cells were designed for remote maintenance (i.e., maintenance

---

[17] The current designation for this type of equipment is "items relied on for safety (IROFS)" as defined in 10 CFR 70.4, "Definitions."

from outside the cell by the use of in-cell cranes, shielding windows, and manipulators). Similarly, a cell was also provided for remote packaging of radioactive wastes and for performing remote decontamination and maintenance of equipment removed from other process cells. The rest of the process cells were designed to permit direct personnel entry and contact maintenance, but only after appropriate remote decontamination has been completed to allow safe entry. These cells were designed to minimize maintenance requirements.

The process and support equipment used in handling radioactive materials is contained in cells or glove boxes. Spent fuel assemblies are stored and transported under water in pools. The cells, glove boxes, and pools provide a barrier between the highly contaminated or radioactive environment within and the habitable environment. Cells with thick concrete shielding walls or pools with deep water cover are provided where protection is required against penetrating (gamma) radiation. Glove boxes are used to isolate radioactive material when radiation levels are low and contact operations are permitted. In the BNFP, the portions of the building allowing personnel access were divided into the radiation zones shown in Table 11. (The historical limits in Table 11 are much higher than the actual radiation fields in modern reprocessing facilities.)

Table 11: Radiation Zones and Permissible Radiation Fields at BNFP

Zone	Radiation Field (maximum)
Normal access, nonradiation zone (area)	0.1 mR/h
Normal access, work zone (station)	1.0 mR/h
Normal access, above work zone (station)	1.0 mR/h (at 1 ft from shield)
Limited access, work zone (gallery)	10 mR/h
Limited access, above work zone (gallery)	100 mR/h (at 1 ft from shield)

The shielding design and designation of each room within the separations facility building are based on the functions to be carried on in the room, the expected occupancy, and the anticipated exposure rate. Personnel access to cells is possible but is allowed only when absolutely necessary and only then with adequate protection and health physics coverage. Cell entry is possible only through heavily shielded doors or hatches, which are normally sealed.

The process equipment, piping, building and structures, casks, storage tanks, and fuel element cladding (prior to shearing) provide barriers for the confinement of radioactive materials. Essential confinement systems are designed to maintain their function under normal operating conditions, abnormal operations, upper limit accident conditions, and adverse environmental conditions throughout the life of the facility. Hatches and penetrations, which are an integral part of the structure, are designed so as not to compromise the confinement and shielding functions.

The floors of all cells in the facility are covered with continuous (welded) stainless steel liners. These liners serve to contain all liquids within the cells in the event of a primary vessel leak. The walls of the cells are covered with either stainless steel or a radiation-resistant paint. The choice of cell wall covering depends on the nature of the material to be processed within the particular cell and the need for decontamination. The wall covering serves to seal the concrete structural material from the corrosive atmosphere and radionuclides and, hence, facilitate decontamination.

Figure 11: BNFP fuel reprocessing plant operating area in front of hot cells [Permission to use this copyrighted material is granted by Allied-General Nuclear Services (AGNS)]

Glove boxes are used to provide confinement when operational requirements and radiation levels permit manual operation. The penetrating radiation produced by the radionuclides within the glove box must be sufficiently low that personnel may operate and maintain the equipment without receiving exposure above approved standards. Therefore, the type of operation performed within glove boxes typically involves only small quantities of radionuclides with penetrating radiation. Generally, glove boxes are used for laboratory, sampling, inspections, or clean plutonium operations. Figure 12 shows a typical glove box setup for handling radioactive material.

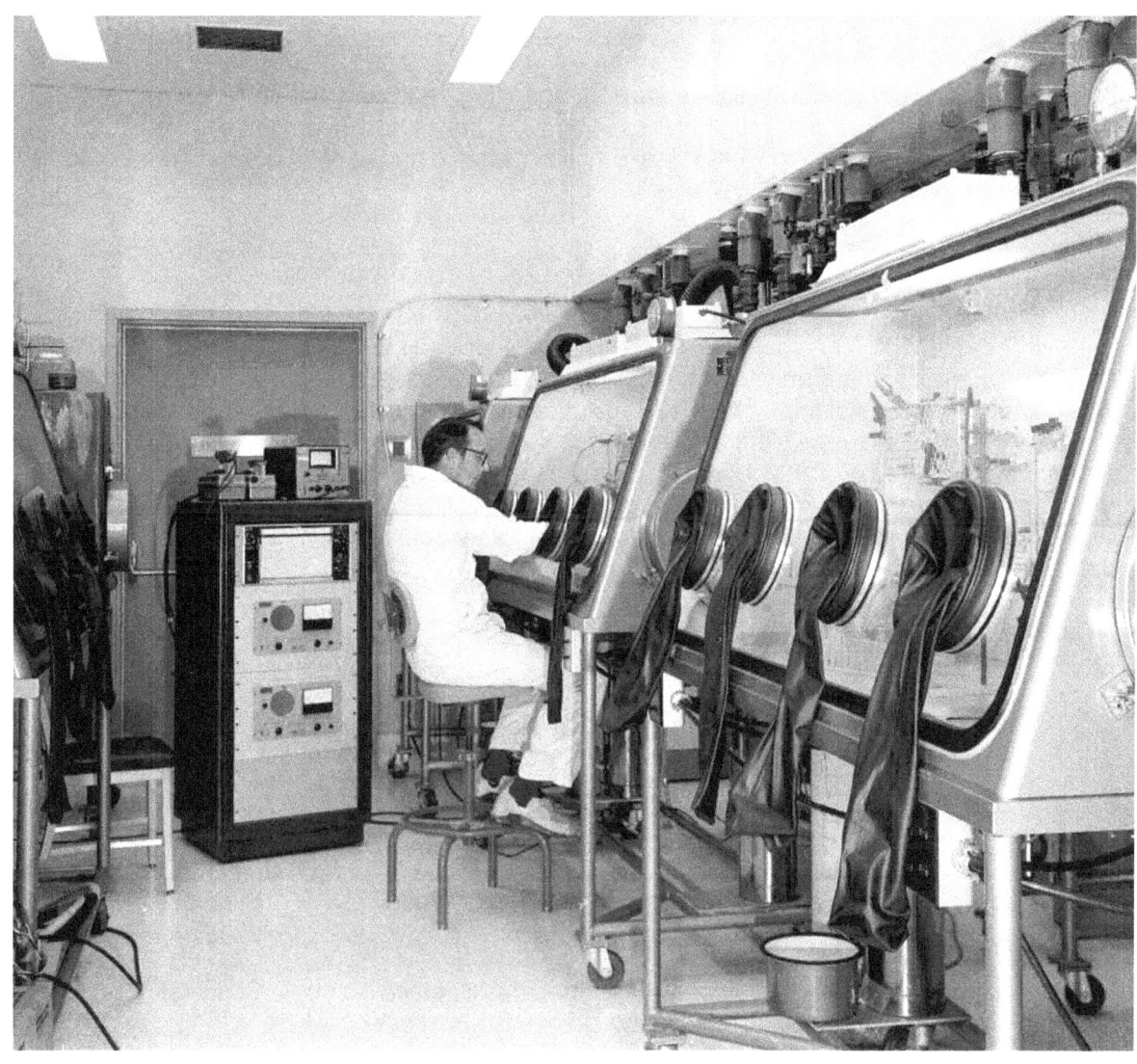

Figure 12: Glove boxes used for handling nuclear materials having low levels of penetrating
radiation

4.2.3. Equipment Modules

The major equipment modules required are (a) spent fuel receiving and storage, (b) main process cells, (c) HLW solidification plant, (d) uranium hexafluoride conversion plant, (e) plutonium product facility, and (f) auxiliary process systems and service areas. These modules are discussed below.

4.2.3.1. Spent Fuel Receiving and Storage

In the BNFP, the fuel receiving and storage station (FRSS) was designed to receive and store wet SNF from LWRs. The spent fuel assemblies are received in shielding casks transported by either truck or rail and are unloaded under water. The fuel assemblies are stored under water to provide cooling and shielding. The FRSS includes facilities for decontaminating the shipping casks before they leave the plant and equipment to circulate, filter, deionize, and cool the spent fuel storage pool water. Table 12 summarizes the major areas of the FRSS and their primary functions. Modern reprocessing plants typically have facilities for unloading dry SNF in air which avoids the need for a water pool and distribution of trace radioactive materials, which requires water cleanup and generates more waste.

Table 12: Primary Functions of Areas in the BNFP Spent Fuel Receiving and Storage Station

Area	Primary Process Functions	Remarks
Two vehicle-loading bays	Receive rail and truck casks; unload casks from transport vehicle; load empty casks onto transport vehicles	
Test and decontamination pit	Prepare casks for unloading in cask unloading pool	Stainless steel floor pan
Cask unloading pools	Remove fuel from casks; place solidified waste canisters in cask	Stainless steel liner
Decontamination pit	Decontaminate casks after removal from cask unloading pool	Stainless steel floor
Fuel storage pool; waste canister racks	Store fuel assemblies and solidified waste	Stainless steel liner
Fuel transfer pool	Transfer fuel assemblies to main process building	Stainless steel liner
Deionization area	Circulate, filter, deionize, and cool pool water	

The FRSS is connected to the main process building by the fuel transfer conveyor tunnel and is connected to the waste solidification plant by an underwater transfer aisle. The pool walls and liners are designed to maintain their containment integrity during a design-basis earthquake or tornado. Building walls above the pools are non-Q structures.

4.2.3.2. Main Process Cells

The main process cells are the functional center of the reprocessing/recycling plant. The uranium and plutonium are chemically separated from the other actinides and fission products in these cells. This processing is carried out in a series of cells that occupy a major portion of the building. The main process building also contains a wide variety of facilities and equipment that is used to monitor and control the process, maintain the equipment, carry out auxiliary operations, and treat gaseous effluents from the processes performed in the building.

Table 13 lists the primary functions of the main process cells. Most of the building is constructed of reinforced concrete designed to remain intact during a design-basis earthquake or tornado. Process cell walls are rebar-reinforced and up to 2 meters thick to provide personnel shielding from radioactivity.

The areas outside the main process cells are generally divided into regions called "galleries," "areas," or "stations." These regions enclose and protect service piping, process support equipment, instrumentation components, and some operating areas. Radioactivity levels range from essentially background to fairly modest levels.

Table 13: Primary Functions of Main Process Cells

Cell	Primary Process Function	Remarks
Remote process cell	Shear and dissolve fuel; concentrate high-level liquid waste	Stainless steel floor pan; remote maintenance
Remote maintenance and scrap cell	Package leached hulls and other solid waste; remotely maintain contaminated equipment	Stainless steel walls and floor
High-level cell	Accountability for dissolver solution; chemically adjust dissolver solution; centrifuge dissolver solution	Stainless steel floor pan
High-intermediate level cell	Separate uranium and plutonium from high-level waste; separate uranium from plutonium; treat dissolver off-gas; clean up solvent; concentrate intermediate-level waste	Stainless steel floor pan; contact maintenance
Intermediate level cell	Treat vessel off-gas; recover nitric acid; concentrate LLW; burn used solvent	Stainless steel floor pan; contact maintenance
Uranium product cell	Purify uranium stream; clean up solvent	Stainless steel pan; contact maintenance
Plutonium product cell	Purify plutonium stream	Stainless steel pan; contact maintenance
Plutonium nitrate storage and load-out	Store plutonium nitrate solutions; transfer plutonium nitrate to plutonium oxide conversion facility	Stainless steel pan; contact maintenance

53

4.2.3.3. Waste Solidification Plant

The waste solidification plant (WSP) is located adjacent to the main process building. It receives high- and intermediate-level liquid waste solutions from the waste tank farm complex, converts the liquids to a vitrified solid, and facilitates the transfer of solidified HLW to the FRSS for storage and eventual shipment off site.

The WSP contains the waste vitrification and canister-sealing equipment, inspection and decontamination equipment, off-gas treatment equipment, and remote maintenance facilities in four process cells. Table 14 presents the primary process functions performed in each of the cells. All process cells in the WSP are completely lined with stainless steel. The cells are surrounded by limited access areas for operating and controlling the processes in the cells. All operational and maintenance facilities in the process cells were to be performed remotely using viewing windows, manipulators, and cranes.

Table 14: <u>Primary Functions of Major Process Cells in the Waste Solidification Plant</u>

<u>Area</u>	<u>Function</u>
Waste vitrification cell	Calcine liquid waste; vitrify calcined waste
Canister decontamination cell	Decontaminate outer surfaces of canisters; transfer filled canisters to FRSS
Off-gas treatment cell	Treat off-gas from WSP process vessels
Hot maintenance cell	Perform remote maintenance on contaminated equipment

4.2.3.4. Uranium Hexafluoride Conversion Plant

In the BNFP, the conversion facility to produce UF_6 consisted of two buildings, both of standard chemical plant construction. The main building was a multistory structure containing the principal process areas. A second building located near the main process area was used for fluorine generation. The UF_6 facility was located near the main process building to eliminate the need for shipping uranyl nitrate to a distant conversion plant. Elimination of the uranyl nitrate shipping requirement saved time, reduced the costs to the nuclear power industry, and lessened the chances of a radiological hazard to the public. Typical UF_6 plants are designed such that there is sufficient surge capacity between process stages to continue operation of adjacent stages when one section is down.

4.2.3.5. Plutonium Product Facility

A plutonium product facility (PPF) was contemplated in the BNFP. Its purposes were to (1) convert aqueous plutonium nitrate solutions to plutonium oxide and (2) provide storage for plutonium oxide. The PPF process facilities were to be located in a separate building immediately adjacent to the main process building.

The PPF was to have a nominal design capacity of 100 kilograms heavy metal of plutonium product per day in the form of plutonium dioxide with an on-stream time of 250 days per year to give an annual conversion capacity of 25,000 kilograms heavy metal of plutonium (1134 grams of PuO_2 contains 1000 grams of plutonium). This capacity would be divided between two identical production lines, each with a capacity of 50 kilograms heavy metal per day. This design basis was selected to allow a 1500 MTIHM per year reprocessing/recycling facility to process MOX feed material for fuels with plutonium contents higher than LWR recycled fuels. Although the design capacity was 25,000 kilograms heavy metal of plutonium per year, the facility could be operated at a lower capacity.

Most of the operations and maintenance in the facility were to be carried out in glove boxes. Leaded gloves were planned to be used to protect against low-level gamma radiation, and relatively thin concrete and other hydrogenous shields were to be used to protect against the neutron radiation from the ^{238}Pu (from (α, n) radiation produced when high-energy alpha particles strike low-atomic number elements).

4.2.3.6. Auxiliary Process Systems and Service Areas

The auxiliary process systems and service areas provide necessary services to the functions of the main process building. The major areas are described below.

4.2.3.6.1. Ventilation System

The ventilation system consists primarily of supply and exhaust subsystems. The ventilation system was designed to provide once-through airflow by pressure controls from non-contaminated areas through potentially contaminated or low-contaminated areas to highly contaminated areas (i.e., process cells), then to treatment systems before being pumped by blowers out the stack. Three blowers were to provide exhaust for the main ventilation system. Each blower was to be capable of supplying 50 percent of the required capacity and was to be connected to emergency power sources.

Exhaust gases from the radioactive processing cells were to pass through at least two stages of high-efficiency particulate air (HEPA) filters. Off-gases from areas with high plutonium concentrations were to pass through three stages of HEPA filters. These extra stages of filtration were designed to provide for a minimum of one stage of filtration in the event of failure of the upstream filters by mechanisms such as fire. Exhaust gases from the main process and building ventilation systems were to exit through the main stack (100 meters high). Non-condensed gases from the concentrator were to vent through the service concentrator stack (30.5 meters high). The chemical makeup and addition tank were to vent through the chemical off-gas stack (29 meters high).

A major feature of the ventilation system design was the ventilation filter station. This housed the primary supply and exhaust blowers and the final stage of HEPA filters through which the air passed before exhausting through the 100-meter stack.[18] This late 20th century design incorporated a cryogenic krypton capture and recovery system, but neither a krypton recovery system nor a system for capture of tritium and ^{14}C was being built when construction ceased.

---

[18] In contrast to the ventilation system planned for BNFP, experience in existing large commercial reprocessing facilities has shown the need for wet scrubbers, condensers, mist eliminators, HEPA filters, etc., to meet effluent discharge limits.

4.2.3.6.2. Electrical Power

From a commercial substation, two transformers, each feeding a 2000-amp main breaker, provided normal electrical power to the facility. The main breakers distributed power through12 120-amp feeder breakers.

The emergency electric power system was designed to handle essential electrical loads in emergency situations. Emergency power was supplied by two independent diesel engine-driven generators. Each generator had a 2200-kilowatt continuous rating. An emergency battery supply was provided for instrumentation in the main control and the waste tank equipment gallery area.

4.2.3.6.3. Fire Protection System

Fire detection and protection systems at the facility were designed to provide early warning and rapid control of fire. Automatic fire detection devices and audible alarms were installed in all areas of the facility. The process cells had dual detection systems. The remotely maintained process cells used manually operated noncombustible purges and water spray mist systems. Automatically operated halon[19] systems served the contact-maintained cells. The filter stations were designed with automatic mist suppression systems, and the FRSS had manually operated fire hoses. Most other areas used a manually operated water sprinkler system.

4.2.3.6.4. Hot and Cold Laboratory Area

The laboratories provided analytical services for all nonradioactive and most radioactive process samples. At the BNFP, the laboratory building was a two-story complex adjacent to the main process building. It was composed of 13 individual laboratories equipped to provide specific types of analysis or services. Radioactive materials in these laboratories were handled in glove boxes. The sample and analytical cells were in a shielded facility designed to facilitate radiochemical analyses of samples from the more highly radioactive portions of the process. The cells provided a shielded area for remote sampling and analysis of these materials and for preparation of samples to be analyzed in the plant analytical laboratories. Operation was to occur through the use of either shielded cells with viewing windows and manipulators or glove boxes.

4.2.3.7. Control Room Area

The control room area housed the process control and safety-related instrumentation for the plant. It served as the communications center from which operators could be directed to perform manual functions. The control room area was not expected to be contaminated under normal operating conditions, since the only process connections to other facility areas were electrical.

---

[19] Halon is a liquefied, compressed halogenated hydrocarbon gas that stops the spread of fire by physically preventing (suffocating) combustion. Although the production of halon in the United States ceased on January 1, 1994, under the Clean Air Act, it is still legal to purchase and use recycled halon and halon fire extinguishers.

4.2.3.8. Liquid Waste Storage Areas

High- and intermediate-level liquid waste from the reprocessing operation would be concentrated and stored in large underground tanks until the wastes could be solidified and shipped off site for disposal. The BNFP had a liquid waste storage complex composed of two high-level liquid waste (HLLW) tanks, one intermediate-level liquid waste (ILLW) tank, and a waste tank equipment gallery (WTEG) to provide services for the tanks. One equivalent HLLW tank volume was to remain available at all times for use as a spare in case of difficulties with any tank of HLLW or ILLW. Additional HLLW tanks were to be added to handle the continued generation of wastes. The WTEG is a concrete building located near the main process building housing the control room, heat exchangers, coolant circulating pumps, off-gas treatment equipment, and ventilation filters for the waste storage tanks. These tanks were connected through a small diverter cell beneath the WTEG and through underground pipe vaults to the main process building and the waste solidification plant. Figure 13 depicts the BNFP plant HLW storage tanks under construction and shows the extensive internal cooling piping required to remove radioactive decay heat. This large amount of coolant piping in storage tanks at plants undergoing decommissioning poses significant problems when solid salts and sludges must be removed, as is the case at some DOE sites. However, at BNFP, all of the tanks were made of stainless steel, which allowed storage of acidic wastes and essentially eliminated the presence of solids. Storing wastes as acidic solutions avoided the formation of sludges (primarily hydroxides and hydrous oxides of metallic ions such as lanthanides, other fission products, and iron) such as those formed in the waste storage tanks at the Hanford and Savannah River sites.

Figure 13: Tanks for liquid HLW storage under construction at the BNFP facility

4.2.3.9. Solid Waste Storage

The BNFP design included a solid waste storage area of approximately 20 acres, an area deemed sufficient to store the solid waste generated during the first 3 years of operation. The solid wastes to be stored in this area were divided into three major categories—(1) spent fuel cladding hulls and hardware; (2) high-level general process trash (HLGPT); and (3) low-level general process trash (LLGPT).[20] Hulls and HLGPT were to be stored in caissons mounted in an engineered berm or in concrete vaults. The LLGPT was to be stored in earth-covered cargo containers. In modern reprocessing plants, the required waste storage volume per unit of SNF processes is likely to be less because of improved operational practices.

4.2.4. Criticality Control Methods

Whenever enriched uranium or plutonium is present, criticality control becomes an important consideration. The method used to control criticality depends on the physical and chemical nature of the fissile material, its mass and purity, and its geometry. Several control methods have been used.

4.2.4.1. Physical Form Control

It is important to know if the physical form is such that fissile material can be compacted to increase its density.

4.2.4.2. Mass Control

For criticality to occur, it is essential that the amount of fissile material equal or exceed the minimum critical mass. A common approach to preventing criticality is to limit the allowable amount of fissile material in any one location to less than a critical mass.

4.2.4.3. Composition Control

Certain chemicals mixed with the fissile material can prevent criticality by absorbing neutrons. Elements with isotopes having large neutron absorption cross-sections such as boron, cadmium, or gadolinium, are commonly added to fissile materials. Usually, these elements are in a form permitting their easy removal when desired.

4.2.4.4. Geometry Control

Vessels having geometries that allow for loss of neutrons through their surfaces in amounts such that a chain reaction cannot be sustained in the vessels are universally used. The vessels may be of many differing configurations, but cylindrical or flat "slab" configurations are common. Typically, one dimension, such as diameter in the case of cylinders or thickness in the case of slabs, is limited to the order of 13 centimeters. Another geometry that has been used is annular tanks, with neutron poisons in the annulus.

---

[20] The NRC does not have a category of waste called "low-level general process trash." BNFP used the term as a descriptive identifier of a type of radioactive waste rather than as a formal waste classification.

4.3. DuPont Reprocessing Studies

After many years of operating the DOE SRS reprocessing plant, the DuPont Company performed research and development (R&D) and supported R&D by others leading to a conceptual design for what would have been an NRC-licensed fuel recycle complex based on DuPont's reprocessing experience and the experience of others. The design studies were completed and reports issued in November 1978 [Kursunoglu, 2000]. White House reviews of reprocessing during the Ford, Carter, and Reagan administrations did not consider this facility design. Many, but not all, of the special features listed below are incorporated in reprocessing plants overseas.

Special features of the DuPont facility design included the following:

- canyon structures for containing process equipment that could be installed, maintained, and replaced remotely using overhead cranes

- use of the best technology available, including centrifugal contactors for the first cycle of solvent extraction, and storage of solutions between process steps

- product recoveries greater than 99.8 percent

- reprocessing of 1-year cooled spent fuel

- personnel access to operating areas, with close control of entry and exit

- vitrification of HLW for ultimate disposal

- flexibility to allow changes, additions, or upgrades of equipment, flowsheets, instruments, etc.

- no accumulation of separated plutonium except in secure surge storage between reprocessing and fuel fabrication

- tritium and krypton capture in addition to iodine capture

- sand filters

- opportunities for lowering cost as a result of longer cooling time before reprocessing

4.4. Operator Licensing and Training

The operation of a reprocessing/recycling facility entails all of the operational skills and safety requirements associated with a reasonably complex chemical processing plant overlain with the radiation safety, security, and safeguards requirements of a nuclear facility. However, other than as noted in the preceding sections of this report, little current commercial experience remains in the NRC-regulated sphere and that which does exist resides mostly in people who have retired. Therefore, the training and qualification of the operating staff takes on major significance in the absence of a pool of fully trained, experienced, and licensed personnel.

In the past, the general criterion was for operators, technicians, and supervisors to have received at least a 2-year certificate from an established technical school. Applicants with this background normally would have sufficient understanding of the physical, chemical, and engineering technologies to undergo the necessary specific plant training.

The importance of qualified operators to the safety of a reprocessing plant can hardly be overemphasized. The regulation in 10 CFR 55.31, "How to Apply," sets forth the contents of an application for licensing individuals who manipulate the controls of a properly licensed facility (at the time of BNFP).

4.4.1. Experience at Nuclear Fuel Services

Experience gained from the licensing of reprocessing plant operators at other commercial reprocessing plants may be of some benefit to this study. This historical experience indicates the validity of the requirement for training and the evolution of training programs over the years, as well as a possible direction for future training efforts.

During the planning stages of NFS, its management and the regulatory staff of the U.S. Atomic Energy Commission (AEC) established four major operator categories:

(1)   manipulator operators
(2)   chemical operators
(3)   control room operators
(4)   senior operators

These categories were similar, in most respects, to those presented in the AEC licensing guide, which was used at that time for nuclear reactor operators.

The results of the original operator examining program in 1966 were disappointing. Of the total number of senior operator applicants taking the examination, 78 percent were successful in obtaining licenses. However, only 59 percent of the chemical operations personnel applying for licenses succeeded, and only 9 percent were initially awarded licenses. Some of the reasons for the excessive failure rate were as follows:

•   At the time of testing, the NFS head-end system had not been completed, and very little practical operating experience could be included in the training program.

•   Most of the applicants were young, and, therefore, had little or no industrial experience.

•   As is usually the case in a new plant, the inadequacies of the first training program were not apparent until the examinations had been completed.

•   The first set of tests was, to some extent, experimental.

The disappointing results and the underlying reasons were similar to those experienced in the early phases of the program for examining power reactor operators.

Later, a pretesting program was conducted at ORNL to establish the validity of future examination procedures. In this program, process foremen, chemical operators, and technicians who had considerable experience in the reprocessing field and new employees with little or no experience took the same tests. The questions posed were basically those to be used for examining NFS operators. The results obtained in this program verified that the questions proposed for the NFS tests were reasonable and confirmed that adequate training was a prerequisite for passing the licensing examination.

As the training methods improved and new testing methods were developed, the number of successful applicants at the NFS facility increased. Table 15 presents a summary of the NFS licensing experience during the period 1966–1970.

Table 15: Experience in Applications Made by, and Licenses Awarded to, NFS Plant Personnel

Year	Initial Examination			First Reexamination	
	Number of Applicants	Licenses Awarded	% of Successful Applicants	Number of Applicants	Licenses Awarded
1966	98	43	44	51	34
1967	30	23	77	2	2
1968	18	16	89	0	0
1969	49	32	65	6	4
1970	23	15	65	6	4
Total or Average	218	129	59	65	44

4.4.2. Experience at the Midwest Fuel Reprocessing Plant

During 1971–1972, the operators of the GE fuel reprocessing plant at Morris, Illinois, underwent formal training to prepare them for licensing. Of the 65 persons included in the program, only 2 failed to qualify for licensing. Many of the candidates for training in the Midwest Fuel Reprocessing Plant had been licensed previously in the NFS plant and had obtained employment with GE when the NFS facility at West Valley, New York, ceased operation. Operators in two general categories were trained for operation in the plant (mechanical processes and remote process equipment). In addition, several senior operators were trained for supervisory roles. It was estimated that more than 220 man-days of effort were expended for each candidate in the training program. The estimated cost for this undertaking, including salary, overhead, and training, was established at $25,000 per individual. This cost would be much higher today, of course, because of inflation.

4.4.3. Experience at Barnwell Nuclear Fuel Plant

BNFP was very nearly completed when U.S. national policy stopped commercial fuel reprocessing. The pre-startup staff of the BNFP included a cadre of operators who had been involved in training and retraining over the previous 1 to 4 years. In addition to the operator training program, programs for others such as analytical laboratory technicians and security patrol officers were also conducted. These programs were necessary to ensure that all operations would be carried out correctly, not only for safety reasons, but also for reasons related to safeguards and physical security.

The operations personnel and analytical technicians at the BNFP were cross-trained. Security officers were also cross-trained in various areas of physical security. As a result, the personnel were considered to be highly trained and knowledgeable in BNFP operations but would have required retraining to deal with any systems modifications to generate a more proliferation-resistant fuel cycle operation.

Operators, technicians, and patrol officers in the various categories did not take the necessary licensing examination to permit operation because BNFP licensing was terminated before completion. However, because of the extensive training and retraining taking place during checkout and "cold-run" operation, it was expected that the failure rate during the licensing examination would be low. The presence of more experienced personnel in any type of operating facility helps reduce the mistakes made by those who, although well trained, are inexperienced.

4.4.4. Training for Operation of the Rokkasho-Mura Reprocessing Plant

About 100 people were trained to operate the Japanese Rokkasho-Mura reprocessing plant through 4 years of hands-on operating experience at the La Hague reprocessing plants.

4.4.5. Typical Reprocessing Plant Operator Training Program

The programs necessary to train reprocessing plant operators are far more rigorous than those employed in conventional industrial chemical facilities. Further, the process of choosing candidates who meet the necessary educational, psychological, and medical requirements to receive this training is a prime concern. The selection of candidates who cannot pass the required certification of licensing examinations results in a financial burden to the enterprise. In addition, unsatisfactorily trained individuals tend to jeopardize safety and hamper efficient operation of the plant under normal as well as abnormal conditions.

Current NRC requirements for training and certification of operators working in the nuclear power industry and in nuclear power plants are found in 10 CFR Part 26, "Fitness for Duty Programs," and 10 CFR Part 54, "Requirements for Renewal of Operating Licenses for Nuclear Power Plants." In addition NRC Form 398, "Personal Qualification Statement—Licensee," gives requirements for manipulating controls of a licensed facility. Appendix D presents additional details on operator licensing.

The qualifications of applicants for operator licenses are determined through two methods of testing: (1) written examinations covering categories such as physics, chemistry, mechanical processing systems, chemical processing systems, equipment and instrumentation, power and auxiliary systems, administrative and procedural rules, and radiological safety and (2) an oral examination.

The time required to adequately train an operator was found to be approximately 1 to 1.5 years.

The qualifications of the operators for future reprocessing/recycling plants are yet to be established, as the role of DOE and the level of its interaction with the NRC and potential commercial owners/operators must still be determined.

4.5. Needed Improvements

One of the cornerstones of the proposed GNEP and closely related AFCI is the development and reduction to practice of SNF separation processes that leave plutonium primarily with actinides other than uranium or fission products. This necessitates equipment and methods for tracking, assay, and accountability of the fissile material content of separations process streams that have not been seen before in this country. The proposed processes will require equipment and detectors for real-time tracking and monitoring and fissile content assay of materials used in fabrication of fuels from fissile material from separation processes.

4.5.1. Improved Processes

Any nuclear fuel recycle plants with improved proliferation-resistance will require precise and accurate tracking, detecting, monitoring, and assaying of the plutonium/low-enriched uranium content of product and waste streams from separation and fabrication processes.

Computer programs to record, evaluate, interpret, and provide real-time output from process equipment and fissile material monitors to local and central monitoring stations are essential for the integrated, large-scale data-handling programs for management of data from all parts of the fuel cycle plant (process control, process monitoring, material transfer, material inventory, portal monitoring) to improve plant proliferation-resistance by interrelating and cross-checking disparate sources of information, as well as to improve plant efficiency. Plant operating parameters should be compared on a continuous basis with computer-simulated normal plant operating parameters to detect, evaluate, and report off-normal operation both locally and remotely as a check on possible illicit operations and improper plant operation. The following sections describe these factors in greater detail.

4.5.2. Improved Equipment

Equipment is required for real-time monitoring and assay of fissile materials in streams containing a mixture of actinides that are to be fabricated without further purification for use in reactors. Equipment for real-time monitoring of spent fuel separation processes, based on recent advances in instrumentation and controls and adaptations of equipment and computerized analysis of data already in use, can possibly improve the tracking of fissile material through the processing steps. Flow rates through pipes and process equipment (e.g., centrifugal contactors, pumps, pulse columns, mixer-settlers, and centrifuges) can be better measured and controlled than in the past. Volume and concentration measurements can be made with greater precision and accuracy in feed and product tanks, thus improving material accountability. Fissile material concentrations and amounts can be measured through better sampling and analysis techniques and subsequent computerized analysis of the data. For example, technology and tools already available can provide more and better radiation energy spectrum measurements and resolution. Flow rates of UF_6 can now be measured accurately. These types of improved measurements make possible the location, identification, and quantification of chemical and isotopic species of interest.

4.5.3. Security and Safeguards

In addition to the normal industrial fences and barriers, nuclear facilities have extra requirements for both physical security and nuclear material safeguards. These two requirements often, but not always, overlap. In light of potential terrorist threats, security and safeguards activities are being stressed, and additional measures are being put into place.

Physical, psychological, and mental requirements of the guard and security forces are specified. These are under continuous review as threat levels are reassessed. Entry portals, coded badges, and other measures are used to control and monitor both personnel and equipment egress and ingress. Internal and external portal monitors are required. Periodic physical inventories of objects containing fissile material are performed.

4.5.4. Detectors

The proposed separation processes will require equipment, processes, and detectors for real-time tracking and monitoring and fissile content assay of materials used in fabrication of fuels and fissile material from the low-decontamination separation processes.

Improvements in the proliferation-resistance of nuclear fuel reprocessing plants through use of more accurate detectors are possible in a variety of areas. A variety of methods for personnel monitoring and recordkeeping of movements and activities of personnel can ensure that there are no illicit activities. Speciation technology (e.g., radiochemical methods for trace concentrations, laser spectroscopy, x-ray absorption fine structure spectroscopy, magnetic resonance techniques, redox speciation, ion-selective electrodes) for materials of interest has improved greatly in recent years [NEA, 1999]. Computerized recording and analysis of data from the sensing and measuring equipment, conducted both locally and at remote locations, permits detection of off-normal operating conditions. This capability is useful both for monitoring plant operations and for maintaining accountability of fissile material. Potential areas of application of some of these new technologies are discussed below.

4.5.5. Material Accountability

As already noted, all nuclear material separation and fuel fabrication processes generate products and wastes that contain fissile material. The amount of fissile material going to waste can be significant for high-throughput processes that operate over relatively long periods of time. If recovered, it could potentially exceed a critical mass, although the fissile material is typically very dilute, and a major effort would be required to recover what was previously deemed to be irrecoverable. Highly sensitive detection and measurement equipment is now available to monitor and assay the plutonium and enriched uranium content of waste streams from separation plants and from both enriched uranium and MOX fuel fabrication processes.

A special accountability problem arises when the minor actinides (neptunium, americium, and curium) are not in secular equilibrium because their concentrations are currently often inferred based on assumed equilibrium. Thus, when secular equilibrium is disturbed by processing, accountability can become much more difficult. This is an important consideration, especially when both plutonium and uranium are present.

Computerized, integrated, large-scale data-handling programs for managing data from all parts of the fuel cycle plant (process control, process monitoring, material transfer, material inventory, portal monitoring) will be a necessary adjunct to any modern reprocessing or fuel fabrication plant. These programs can greatly improve plant proliferation-resistance by interrelating and cross-checking disparate sources of information.

Considerable effort, both nationally and internationally, is required among the groups responsible for establishing the permissible significant plutonium inventory differences (Sigma ID).

As shown in Table 16 [Pasamehmetoglu, 2006], there is a large difference among the IAEA, the NRC, and DOE with regard to the sigma ID requirements and the frequency of both long-term shutdown inventory and interim inventory requirements.

In large complex facilities requiring many measurements errors are combined to determine the uncertainty in ID. The ID uncertainty determines the required capability of the safeguards system to detect loss.

The IAEA Sigma ID value is an absolute value of 2.42 kg, independent of facility throughput. NRC and DOE Sigma ID requirements are percentages of the active inventory so their values change with throughput. Table 16 gives estimated values of Sigma ID for the three agencies for a plant with a yearly throughput of 2500 MTHM.

Table 16: Sigma ID Goals for IAEA, NRC and DOE

LWR spent fuel processed yearly	2500 MTIHM
Pu processed yearly (1% of plant throughput)	25,000 kg
Pu processed per month	2,083 kg
IAEA goal	2.42 kg
NRC goal	2.083 kg
DOE goal	20.83 kg

For recycle facilities to be commercially viable, attaining the NRC and IAEA Sigma ID is a political, diplomatic, and technological challenge.

In general safeguards systems are intended to meet certain design objectives for facility operations, and nuclear material transportation. For facilities, the objectives include but are not limited to:

(1)   Ensure that only authorized personnel and materials are admitted into material access areas (MAAs) and vital areas (VAs).

(2)   Ensure that only authorized activities and conditions occur within protected areas, MAAs and VAs.

(3)   Ensure that only authorized movement and placement of source and special nuclear material (SSNM) occur within MAAs.

(4)   Ensure that only authorized and confirmed forms and amounts of SSNM are removed from MAAs.

(5)   Ensure timely detection of unauthorized entry into protected areas.

(6)   Ensure that the response to any unauthorized activity is timely, effective, and appropriate to the particular contingency.

(7)   Ensure the presence of all SSNM in the plant by location and quantity.

For nuclear material transportation, the objectives include:

(1)   Restrict access to and personnel activity in the vicinity of transports.

(2)   Prevent unauthorized entry into transports or unauthorized removal of SSNM from transports.

(3)   Ensure that the response to any unauthorized attempt to enter vehicles and remove materials is timely, effective, and appropriate for the particular contingency.

In general, organizations should always consider the potential for improving overall safeguards performance or reducing the overall societal impacts attributable to safeguards.

The NRC's safeguards program for commercial licensees is part of a national safeguards structure introduced initially to protect defense-related SSNM. The structure includes three primary components: (1) intelligence gathering, (2) site and transportation security, and (3) recovery of lost material. Only the second component, site and transportation security, which involves physical security and material control, would fall primarily within the NRC's field of responsibility. The other two, intelligence and recovery operations, remain the responsibility of other agencies such as the Federal Bureau of Investigation, the National Security Council, DOE, and State and local law enforcement agencies. The NRC collaborates with these other agencies in developing contingency plans for reacting to and dealing with theft or diversion but does not participate in intelligence operations or physically take part in recovery operations.

5. OVERVIEW OF ADVANCED SPENT NUCLEAR FUEL RECYCLE INITIATIVES

The National Energy Policy [NEP, 2001], issued by President Bush in May 2001, recommended expanded use of nuclear energy in the United States, including development of advanced nuclear fuel cycles, reprocessing, and fuel treatment technologies. Consistent with the President's policy, DOE adopted a strategy involving four facets—Nuclear Power 2010; AFCI; Generation IV Nuclear Energy Systems; and the nuclear hydrogen initiative. Additionally, on February 6, 2006, the Secretary of Energy launched GNEP, a comprehensive international strategy to expand the safe use of nuclear power around the world.

5.1. Advanced Fuel Cycle Initiative

The purpose of the DOE AFCI program is to develop fuel systems and enabling fuel cycle technologies for Generation IV (GEN IV) reactors and future reactors in support of GNEP. DOE anticipates that AFCI will provide options for the management of SNF through treatment and transmutation of radionuclides that will reduce the cost, hazards, and volume of HLW disposal in repositories, reduce the amount of plutonium accumulating in the nuclear fuel cycle, and recover for beneficial use the energy potential remaining in spent fuel. DOE plans call for systems analysis to be an important part of the ongoing AFCI program and to have an increased role during the next few years. The planned systems analysis will investigate key issues such as the required rate of introduction of ABRs and actinide separation facilities to avoid the need for a second HLW repository early in this century and a detailed study of the technical requirements for the facilities and how the facilities might support the main goals of the program. DOE plans to use the results of these analyses to establish the basis for each key decision in the AFCI program and for GNEP program planning.

AFCI is organized into the following program elements:

- separations
- fuels
- transmutation
- university programs

The purpose of each element is summarized below.

5.1.1. Separation

Separation processes will be devised to recover plutonium in such a way such that it is never separated from at least some TRU actinides and possibly some fission products. Essentially all of the TRU elements, in addition to the ^{137}Cs and ^{90}Sr, will be removed from the waste going to the geologic repository. Such removal would reduce the heat load in the repository, greatly increasing the number of fuel assemblies whose wastes go to the repository, and consequently obviating the need for additional repository space for many decades. These separation technologies are not alternatives to a geologic repository but could help reduce the cost and extend the life of a geologic repository.

5.1.2. Fuels

Fuel forms for advanced fast-spectrum transmutation reactors that are planned for transmuting TRU actinides (i.e., neptunium, plutonium, americium, and curium) to fission products are being developed. Oxide, nitride, carbide, and metallic fuels are being considered. The AFCI is also developing fuels for GEN IV power reactors.

5.1.3. Transmutation

Transmutation is a process by which long-lived radioactive isotopes, especially actinides such as plutonium and neptunium but also selected fission products such as ^{99}Tc and ^{129}I, are converted to shorter-lived fission products or stable isotopes by fission and/or neutron capture from neutrons generated in a reactor or by the interaction of high-energy ions from a particle accelerator with a metal target such as mercury, tungsten, or bismuth. Theoretically, the preferred neutron source to fission actinides is one of high average neutron energy (yielding a high neutron fission-to-capture ratio), high flux (to which the transmutation rate is proportional), and large core volume (to accommodate more actinides). This has led to a preference for fast reactors as the neutron source. Transmutation of fission products is usually more efficient in the low-energy neutron spectrum typical of thermal reactors such as LWRs, but DOE is currently focusing on actinide transmutation and, thus, on development of fast reactors with the lead candidate being a sodium-cooled reactor with stainless-steel-clad fuel.

5.1.4. University Programs

The goal of the AFCI university programs is to foster education of the next generation of scientists and engineers who will support the growth of nuclear power. This goal is to be achieved primarily by funding infrastructure upgrades at universities and by education and research.

5.1.4.1. University Nuclear Infrastructure

This program brings together several program elements supporting the increasingly vital university nuclear engineering infrastructure. Program elements include the following:

- Innovations in Nuclear Infrastructure and Education: This program strengthens the Nation's university nuclear engineering education programs through innovative use of the university research and training reactors and encouraging strategic partnerships among the universities, the DOE national laboratories, and U.S. industry. Currently under this program, six university consortia provide support for 38 universities in 26 States.

- Reactor Fuel Assistance: DOE provides fresh fuel to, and takes back spent fuel from, university research reactors. Currently, 27 university research reactors are operating at 26 institutions in the United States.

- Reactor Upgrades: DOE provides assistance to universities to improve the operational and experimental capabilities of their research reactors. The universities receive grants to purchase equipment and services necessary to upgrade the reactor facilities, such as reactor instrumentation and control equipment; data-recording devices; radiation, security, and air-monitoring equipment; and gamma spectroscopy hardware and software.

- Reactor Sharing: Through this assistance effort, DOE enables universities with reactors to "share" access to their facilities with students and faculty at other institutions who lack such a facility. The reactors are made available for use in research, experiments, material irradiations, neutron activation analysis, training, and for facility tours and other educational activities.

5.1.4.2.   Nuclear Engineering Education Research Grants

This highly competitive, peer-reviewed program provides grants to nuclear engineering faculty and students for innovative research in nuclear engineering and related areas. The awards run from 1 to 3 years and are granted in nine separate technical areas related to reactor physics, reactor engineering, reactor materials research, radiological engineering, radioactive waste management, applied radiation science, nuclear safety and risk analysis, innovative technologies, and health physics.

5.1.4.3.   Other University Support Activities

These activities include the following:

- DOE/Industry Matching Grants: DOE and participating companies provide matching funds of up to $60,000 each to universities for use in funding scholarships, improving nuclear engineering and science curricula, and modernizing experimental and instructional facilities. The program provides nuclear engineering/health physics fellowships and scholarships to nuclear science and engineering programs at universities.

- Radiochemistry: DOE awards 3-year grants to support education activities in the field of radiochemistry in the United States. Radiochemistry is linked to several national priorities including medicine, energy, and national defense.

- Nuclear Engineering and Science Education Recruitment Program: This program is designed to increase the number of students entering a university nuclear engineering course of study by developing a core curriculum to instruct high school science teachers in nuclear science and engineering topics through the use of teaching modules, teacher workshops, and other outreach activities.

- Summer Internships at National Laboratories: The Office of Nuclear Energy offers summer internships at the Idaho National Laboratory (INL) in technical areas related to nuclear engineering to undergraduate and graduate students.

- International Student Exchange Program: This program sponsors U.S. students studying nuclear engineering for 3–4 months abroad to do research at nuclear facilities in Germany, France, and Japan. These three countries send their students to the United States for reciprocal internships at DOE national laboratories.

5.2. Global Nuclear Energy Partnership (GNEP)

GNEP is a broad-scope DOE program with the goal of promoting beneficial international uses of nuclear energy through a multifaceted approach. Many of the ideas explored earlier by INFCE are embodied in GNEP, which is essentially an updated expression and extension of those ideas.

DOE has entered a Notice of Intent to Prepare a Programmatic Environmental Impact Statement (PEIS) for the Global Nuclear Energy Partnership in the *Federal Register* [DOE, 2007]. This notice gives details of the expected content of the PEIS as well as considerable information about the DOE concept of GNEP.

5.2.1. GNEP Goals

GNEP continues to evolve in response to new information, new international alliances, and changing program leadership. The general goals of GNEP as expressed by DOE in its strategic plan [GNEP, 2007a] are as follows:

> The United States will build the Global Nuclear Energy Partnership to work with other nations to develop and deploy advanced nuclear recycling and reactor technologies. This initiative will help provide reliable, emission-free energy with less of the waste burden of older technologies and without making available separated plutonium that could be used by rogue states or terrorists for nuclear weapons. These new technologies will make possible a dramatic expansion of safe, clean nuclear energy to help meet the growing global energy demand.

DOE plans three facilities to implement GNEP:

(1)   an industrial-scale nuclear fuel recycling center (Consolidated Fuel Treatment Center [CFTC]) to separate the components of spent fuel required by GNEP

(2)   an advanced burner reactor (ABR) to fission the actinides yielding fission products that are more readily managed while producing electricity (DOE is leaning toward a sodium-cooled fast reactor for the ABR)

(3)   an advanced fuel cycle research facility (Advanced Fuel Cycle Facility [AFCF]) to serve as an R&D center of excellence for developing transmutation fuels and improving fuel cycle technology

Two approaches are being used to develop these three facilities. Industry, with technology support from laboratories, international partners, and universities, would lead the development of the CFTC and the ABR. The AFCF would be located at a Government site, DOE would fund the research at the facility, and the national laboratories would take the lead in creating the technology used in the CFTC and fuels for the ABR.

DOE expects the components of GNEP to provide the following benefits:

•   expand domestic use of nuclear power and reduce dependence on fossil fuels

- demonstrate more proliferation-resistant fuel recycle processes

- minimize high-heat-output nuclear waste and thus obviate the need for additional U.S. geologic repositories before 2100

- develop and demonstrate ABRs to produce energy from recycled fuel

- establish reliable fuel services to participating nations by providing fuel on a lease-and-return basis

- demonstrate small-scale reactors

- develop enhanced nuclear safeguards by designing safeguards directly into nuclear facilities and reactors and by enhancing IAEA safeguards capabilities

5.2.2. GNEP Timetable—Phased Approach

On August 3, 2006, DOE announced $20 million for GNEP siting studies and sought further cooperation with industry through issuance of a Request for Expressions of Interest in licensing and building a CFTC and an ABR.

The GNEP is a phased program. Each phase begins after a decision based on the results of the previous phase and an assessment of the risks associated with proceeding to the next phase. DOE has stated that it will proceed to detailed design and construction of the GNEP facilities after it is confident that the cost and schedules are understood and after the project management framework that will allow these projects to succeed is in place.

It is anticipated that the NRC will regulate the CFTC and ABR. The AFCF will be built on a DOE site and is not expected to be licensed by the NRC. Because the GNEP policy and technological approaches to implementing the policy continue to change, it is important that the NRC have a strategy to accommodate the changes, both in allocation of personnel and budgeting.

5.3.   Russian "Equivalent" Proposal (Global Nuclear Infrastructure)

Russian President Putin put forward in 2006 a broad nonproliferation initiative called the Global Nuclear Infrastructure (GNI), which envisions the establishment of international nuclear centers, and offered to host the first such center in Russia. The proposed centers would provide participating nations with full "nuclear fuel cycle services," including enriching uranium, fabricating fresh nuclear fuel, and storing and reprocessing SNF.

Under the terms of the Nuclear Non-Proliferation Treaty, states (nations) not possessing nuclear weapons are permitted to engage in uranium enrichment and spent fuel reprocessing, but these activities are considered to pose significant proliferation risks because they can provide access to weapons-usable nuclear material. The Russian nuclear center proposal would concentrate such activities in states already possessing nuclear weapons and would limit the introduction of enrichment and reprocessing facilities in states without nuclear weapons.

Russia has stated that it would be ready to set up a pilot international enrichment center. This center would provide non-weapons nuclear power states with assured supplies of low-enriched uranium for power reactors and would give them equity in the project without allowing them access to the enrichment technology. The existing uranium enrichment plant at Angarsk, the smallest of three Siberian plants, will become part of the international center which will be under IAEA supervision. The enriched uranium will be subject to safeguards. Russian legislation is needed to separate the facility from the defense sector and open it to international inspection, as well as to provide for a shareholding structure for other countries involved with the center.

GNI will be the first expression of President Putin's initiative which is in line with the IAEA 2003 proposal for multilateral approaches to the nuclear fuel cycle. GNEP proposals involving such centers are very similar [WNA, 2006], and collaboration with the Russian initiative is anticipated.

5.4. Generation IV Nuclear Reactors

The Generation IV International Forum was chartered in May 2001 to lead the collaboration of the world's eminent nuclear technology nations to develop next-generation nuclear energy systems (reactors) to meet the world's future energy needs. This international effort reached a major milestone on February 28, 2005, when five of the forum's member countries signed the world's first agreement aimed at the international development of advanced nuclear energy systems.

The forum identified five distinctly different reactor systems for development [NERAC, 2002]. Initial emphasis was to be placed on those reactors whose next generation would be evolutionary improvements of PWRs and BWRs, rather than radical departures from existing technology. All five of the reactor systems have operating experience (PWR, BWR, sodium fast reactors, and HTGR) or extensive research and development (MSR)

5.5. Nuclear Power 2010

The technology focus of the Nuclear Power 2010 program is on Generation III+ advanced LWR designs which offer advances in safety and economics over the Generation III designs licensed by the NRC in the 1990s. To enable the deployment of new Generation III+ nuclear power plants in the United States in the relatively near term, it is essential to complete the first-of-a-kind Generation III+ reactor technology development and to demonstrate the use of untested Federal regulatory and licensing processes for the siting, construction, and operation of new nuclear plants. DOE has initiated cooperative projects with industry to obtain NRC approval of sites for construction of new nuclear power plants under the early site permit process, to develop application preparation guidance for the combined construction and operating license (COL), to resolve generic COL regulatory issues, and to obtain NRC approval of COL applications. The COL process is a one-step licensing process by which public health and safety concerns related to nuclear plants are resolved before construction begins and the NRC approves and issues a license to build and operate a new nuclear power plant. Utilities have begun to apply for new reactor construction and operating licenses.

Although DOE is supporting industrial development of improved and advanced reactor designs, few if any new reactor construction starts will occur before 2010. However, there have been a substantial number of operating license renewal applications. As of August 2007, the NRC had received license renewal applications for 57 reactor units and had approved 20-year license extensions for 48 reactor units.

6.   ADVANCED FUEL REPROCESSING TECHNOLOGY

In the early years of reprocessing in the United States, the goal was to separate pure plutonium containing a high proportion of ^{239}Pu for use in nuclear weapons. Irradiations in the plutonium production reactors at the DOE Hanford and Savannah River sites were carried out for short times to minimize the generation of undesirable higher mass number plutonium isotopes. As interest in commercial power-producing reactors grew, the emphasis changed from weapons plutonium production operating conditions to higher fuel burnups to maximize energy production and minimize cost. This emphasis led to a smaller proportion of weapons-grade ^{239}Pu and larger proportions of ^{240}Pu, ^{241}Pu, and ^{242}Pu in the spent fuel.

The ongoing DOE reprocessing development program focuses on proliferation-resistant processes. DOE's preferred approach to increasing proliferation-resistance is to eliminate the pure plutonium product. Other important goals for future reprocessing plants include minimizing the volume of radioactive wastes produced by the plant, decreasing losses of fissile and fertile elements to waste (most notably plutonium and uranium), and removing heat-producing radionuclides in from the HLW. As shown in Figure B1 in Appendix B, radionuclides constituting the major source of decay heat in SNF are ^{137}Cs and ^{90}Sr in the medium term and the actinides, primarily plutonium and ^{241}Am , in the long term. This fact is the impetus for actinide removal in the UREX processes. However, many of the UREX processes under development by DOE are not yet optimized with respect to minimizing the number of separation cycles or achieving the requisite separation efficiencies. The OECD Nuclear Energy Agency has generically evaluated once-through, partially closed, and fully closed fuel cycles against multiple criteria [NEA, 2006].

It is important to know the efficiencies of the separation processes used in the flowsheets. This information is obtained as nearly as possible through laboratory experiments with nonradioactive materials, followed by experiments with radioactive tracers, then with small amounts of irradiated fuel, and finally by small-scale integrated process experiments with irradiated fuel. At the same time, the various pieces of process equipment are tested individually and then as integrated systems to ensure that process goals will be met. These latter tests may be performed without using radioactive material, or with uranium only. Data from the laboratory and equipment tests are used to select and design pilot plant recycle facilities. These tests also yield data on separation factors,[21] which are a measure of separation efficiencies for the suite of elements of interest. Besides data for uranium and plutonium, separation data on cesium, strontium, technetium, iodine, neptunium, americium, curium, and the lanthanide elements are very important because the extent of separation determines the distribution of these radionuclides among the products and waste streams and thus determines the need for additional cleanup or helps define disposal routes. Radionuclides previously considered to be of little importance that may be significant in the future include tritium, ^{85}Kr, and ^{14}C.

Because the power densities and fuel burnups in commercial power reactors have been increasing steadily as better information on reactor and fuel performance has become available, and because the half-lives of the radioisotopes cover an enormous range, it is very important to know how much of each radioisotope is produced and how long they are permitted to decay before designing the process or processes to be used in a reprocessing facility and the degree of

---

[21] Separation factor is defined as the concentration of the species of interest in the feed to one stage of the separation process divided by its concentration in the product of that stage of the separation process.

separation required. Decay time is of particular importance in the case of ^{241}Am, most of which grows in as a result of ^{241}Pu decay after the fuel is removed from the reactor.

All of this information goes into establishing mass balance and equipment flowsheets. With such a wide range of variables (fuel burnup, reactor neutron flux, radioactive decay, many radionuclides, degrees of separation for individual radioisotopes or groups of radioisotopes, and equipment options), the number of possible flowsheets becomes very large. Considerations such as degrees of separation sought, process simplicity, ease of process operation, cost, volume of wastes generated, safety, regulations, criticality, and proliferation-resistance of the processes are helpful in selecting the processes that are actually worthy of study and adoption.

6.1. UREX Processes

GNEP has conceived of a suite of UREX processes, each of which consists of a series of steps designed to remove specific groups of radionuclides to tailor products and compositions of the desired product and waste streams [Laidler, 2006]. The PUREX process can be modified readily to be the first step of any of the UREX processes. This step is followed by processes to remove major heat-producing radionuclides from wastes going to the repository and to aggregate TRU actinides for recycle. Table 17 identifies several UREX variants. The variants involve increasing fractionation of the spent fuel constituents as the number of the variant increases.

Table 17: Variants of the UREX Process

Variant Number	Prod # 1	Prod # 2	Prod # 3	Prod # 4	Prod # 5	Prod # 6	Prod # 7
UREX+1	U	Tc	Cs/Sr	TRU+Ln	FP except Cs, Sr, Tc, Ln		
UREX+1a	U	Tc	Cs/Sr	TRU	FP except Cs, Sr, Tc		
UREX+2	U	Tc	Cs/Sr	Pu+Np	Am+Cm+Ln	FP except Cs, Sr, Tc, Ln	
UREX+3	U	Tc	Cs/Sr	Pu+Np	Am+Cm	FP except Cs, Sr, Tc	
UREX+4	U	Tc	Cs/Sr	Pu+Np	Am	Cm	FP except Cs, Sr, Tc

NOTES: TRU = Transuranic elements: Np, Pu, Am, Cm
       FP = Fission products
       Ln = Lanthanide fission products: elements 57 (lanthanum) through 71 (lutetium)

The DOE has been focusing on the UREX+1a flowsheet which produces fissile material products that contain separated uranium in one stream and all the TRU actinides in another. The TRU actinides are to be fabricated into reactor fuel for transmutation and energy. Recently, DOE's interest has been increasing in the UREX+2 flowsheet, which separates americium/curium/lanthanides from the plutonium/neptunium, and the UREX+3 process, which separates the lanthanides from the americium/curium/lanthanide mixture produced by UREX+2.

One objective of the UREX processes is to increase the proliferation-resistance of fuel recycle by avoiding the production of a pure plutonium stream and to fission plutonium and the other actinides to produce energy. A second objective is to remove the major sources of decay heat that limit the spacing of waste packages in a geologic repository. Figure B1 in Appendix B shows that the heat production rate of the actinides exceeds that of the fission products after about 70 years, which illustrates the advantage of removing them from the waste sent to the repository. There is also a potential advantage in keeping the lanthanides with the actinides from the point of view of proliferation-resistance. Disadvantages from keeping the lanthanides with the actides during transmutation remain to be evaluated.

The first UREX process step is a modification of the conventional PUREX process in which the plutonium is prevented from being extracted with the uranium. Plutonium extraction is prevented by chemically reducing extractable Pu(IV) that is normally extracted in PUREX to in-extractable Pu(III) in the first extraction cycle using, for example, acetohydroxamic acid (AHA). Leaving the plutonium combined with other actinides and fission products is believed to provide greater proliferation-resistance than the PUREX process, wherein the plutonium is extracted with the uranium and subsequently separated from uranium and further purified. *It should be observed that a relatively simple change in the first UREX process step (failure to add the Pu(IV) reductant) would result in co-extraction of uranium and plutonium, which would be essentially the PUREX process.* The AHA also reduces neptunium so that it accompanies the other TRU elements. Section 3.1.3, which discusses THORP process chemistry, addresses this point in greater detail.

All UREX variants remove dissolved ^{99}Tc ($t_{1/2}$ = 2.12x10^5 yr), whose most common chemical species under oxidizing conditions is the environmentally mobile pertechnetate anion (TcO$_4^-$), and the relatively short-lived, high-heat-producing fission products ^{137}Cs ($t_{1/2}$ = 30 yr) and ^{90}Sr ($t_{1/2}$ = 29 yr) from the fission product waste stream. The UREX+1a variant routes all the TRU elements and possibly some low-enriched uranium into a single product stream for recycle to the transmutation (burner) reactor. In the transmutation reactor, the TRU elements would be fissioned to produce energy and what is primarily a fission product waste, thus removing by transmutation the principal long-term heat-producing actinides from the wastes.

As of early 2007, some UREX+1a experiments with irradiated fuel had been performed, but no engineering-scale demonstrations have occurred. The difficulties associated with continuously operating any of the UREX variants have not yet been addressed. These difficulties are likely to pose serious operational challenges as all UREX variants require multiple processes operating sequentially, use of multiple extractants, different types of equipment, and multiple solvent cleanup and recycle processes. The staff operating such a plant will require extensive and expensive training. Additionally, if one of the separation process steps were to become inoperable, the entire plant would be shut down because the individual processes must operate sequentially and simultaneously unless the plant has substantial surge capacity between processes.

Flowsheet and process development is underway at ANL, INL, SRS, and ORNL in hot cells at the bench-top scale and at the kilogram scale to establish the viability of the various separation processes. This work, especially sequential kilogram-scale process operation in the hot cells, is very important for establishing the feasibility and performance of the UREX flowsheet. It will be necessary to accompany process development with engineering-scale testing of major equipment pieces and processes.

The UREX+3 variant is noteworthy because it yields a mixture of americium and curium as a product stream separate from the neptunium and plutonium. This feature may be important, depending on how the approximately 55,000 MTIHM of long-cooled spent fuel currently in storage at the reactor sites or spent fuel storage sites are phased into the reprocessing plant processing schedule along with the 2200 MTIHM of spent fuel being generated annually from the existing 104 commercial power reactors (plus the fuel from any new reactors that come on line). Because of radioactive decay and their nuclear properties, the americium and curium from spent fuel aged 35 to 40 years is more efficiently burned in LWRs than in fast reactors [ORNL, 2007], a fact that has the potential to reduce the number of or to eliminate the need for fast reactors currently planned for transmutation of actinides to fission products.[22]

Figure 14, is a block diagram of the UREX+1a process, shows the process steps as the head-end, central, and tail-end unit operations. The head-end operations include chopping of the fuel elements into small pieces, fuel dissolution, and feed clarification to provide the input stream (H--5) to the central UREX+1a process. Additional head-end process steps will likely also include trapping and immobilizing the gases ^{85}Kr, ^{129}I, ^{14}CO$_2$ and ^{3}H. In addition, the hardware and hulls are shown to be compacted and packaged for disposal. These additional head-end steps are, with the exception of iodine retention and hardware and hull compaction, not current reprocessing practice. Figures 15, 16, 17, and 18 show more details on the four major processes in this flowsheet. Figures 14 through 17 were prepared by the authors based on information provided in papers and presentations given by ANL staff members [Periera 2005, 2007] describing bench-scale testing of UREX flowsheets and general considerations related to the design of full-scale reprocessing plants [Benedict, 1981; Long, 1978] such as the need for process steps to clean impurities from the solvent and allow it to be internally recycled.

The four central unit operation steps (UREX, CCD-PEG, TRUEX, and TALSPEAK) are summarized as follows:

- UREX: The uranium and technetium are separated from the dissolver solution fed to this process step and then the technetium is removed by ion exchange. The uranium (uranyl nitrate) product stream undergoes denitration and solidification and packaging for storage. The technetium is converted to metal for disposal, presumably with the fuel cladding hulls.

---

[22] The results in ORNL/TM-2007/24 led the authors to conclude, "Because the ABR design has been optimized at ~840 Mwt, a large number (33–90) of ABRs would be required to transmute the ~23 MT/year TRU actinides currently produced in ~2000 MT/year of low-enriched uranium spent fuel; in comparison, 10–24 existing (or new) 3400 Mwt LWRs would be sufficient" [ORNL, 2007].

- CCD-PEG: ^{137}Cs and ^{90}Sr are separated from the UREX raffinate and stored as glass-bonded aluminosilicates after immobilization by steam reforming.

- TRUEX: The remaining fission products other than the lanthanides are separated from the CCD-PEG raffinate, combined with other waste streams, vitrified, and sent to interim storage.

- TALSPEAK: The TRU elements in the TRUEX product are separated from the lanthanides. The TRU element product from TALSPEAK may be blended with uranium for calcination, packaging and interim storage pending refabrication into transmutation reactor fuel. The lanthanides are combined with the other fission products for vitrification.

The waste forms and waste management strategy outlined above should be regarded as provisional. DOE is preparing a waste management strategy [Wigeland, 2007] to better define the wastes resulting from UREX. The four central process operations in the UREX+1a flowsheet are discussed in detail below.

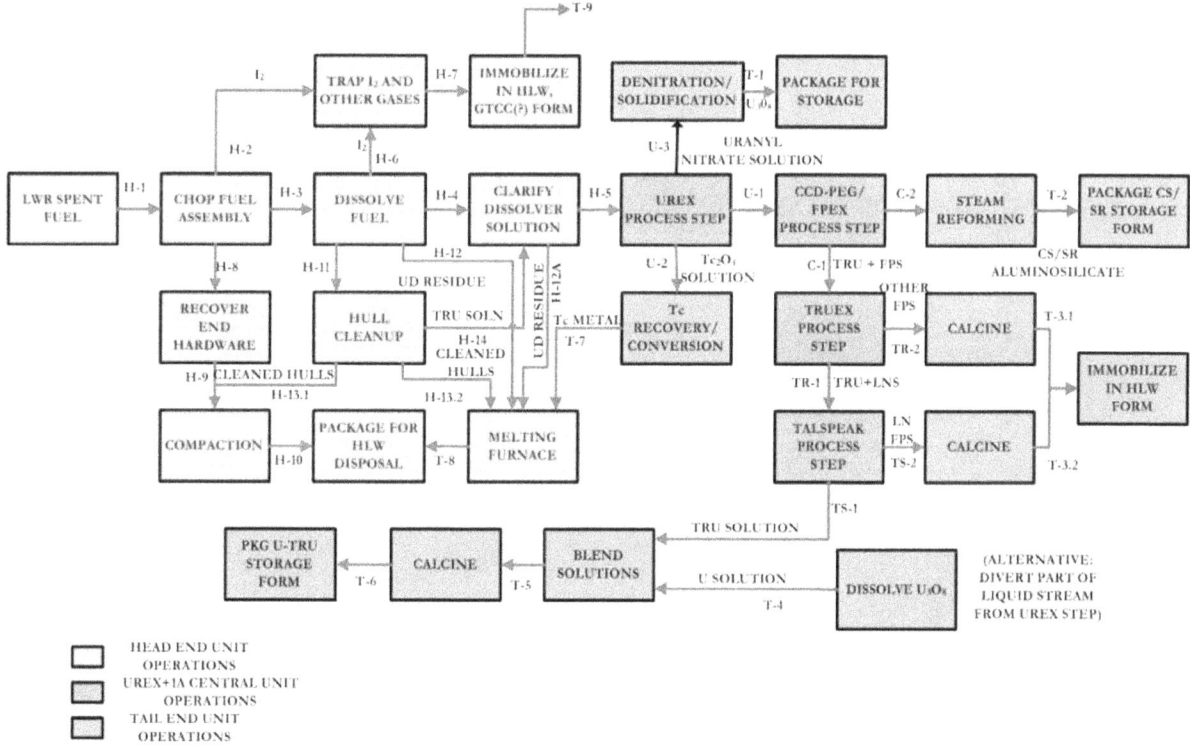

Figure 14: Diagram of primary UREX+1a flowsheet unit operations

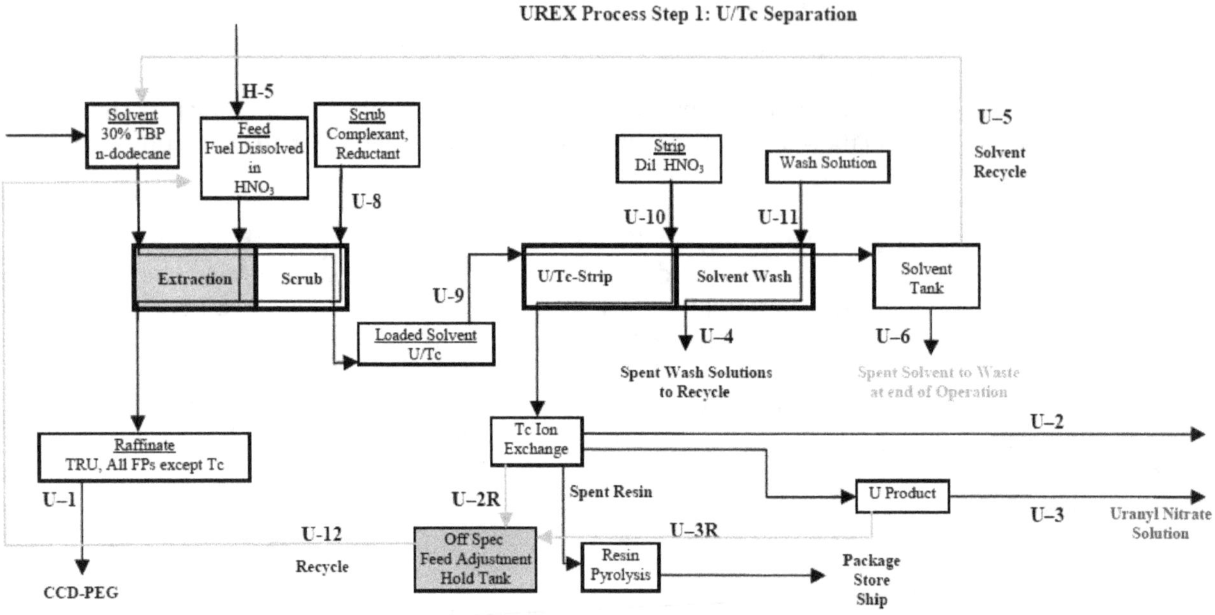

Figure 15: Diagram of UREX+1a Step 1 UREX (modified PUREX) to separate uranium and technetium.

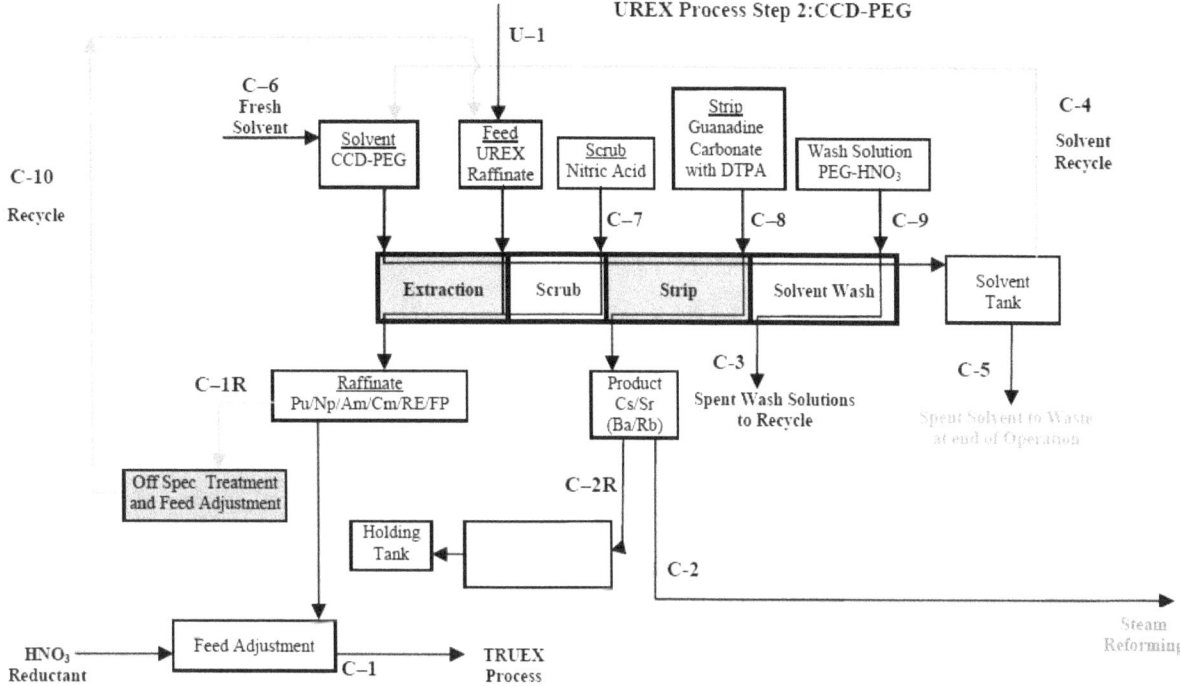

Figure 16: Diagram of UREX+1a Step 2 CCD-PEG to remove cesium/strontium

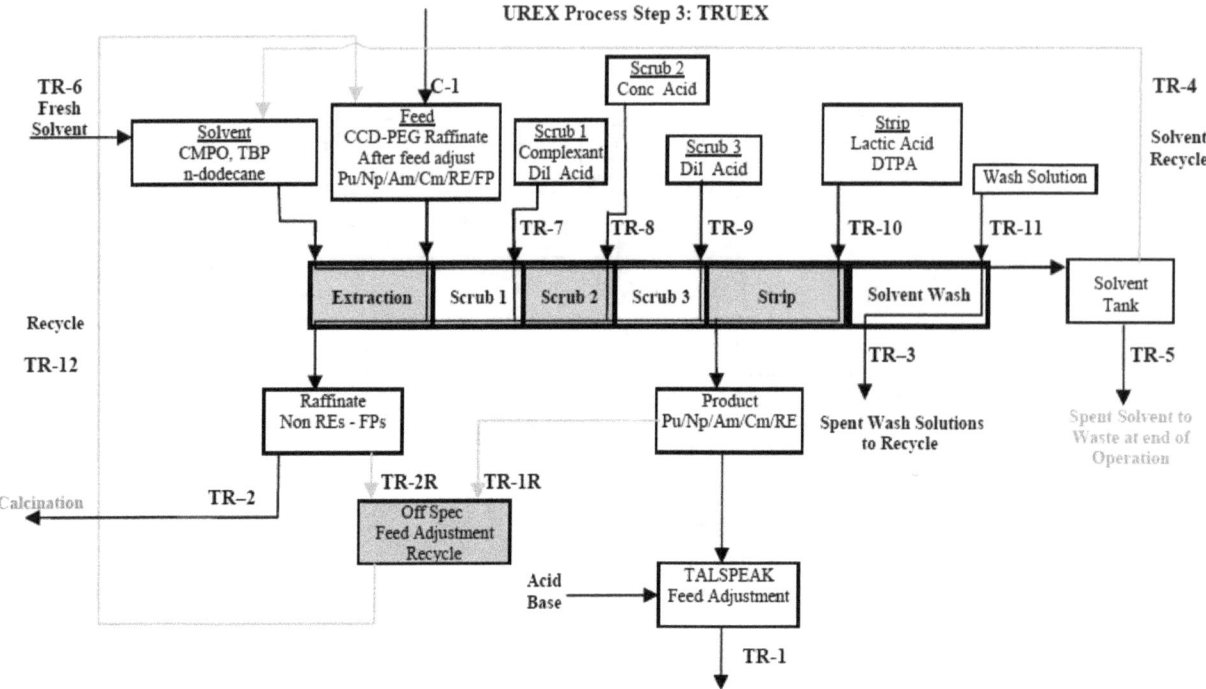

Figure 17: Diagram of UREX+1a Step 3 TRUEX to remove nonlanthanide fission products

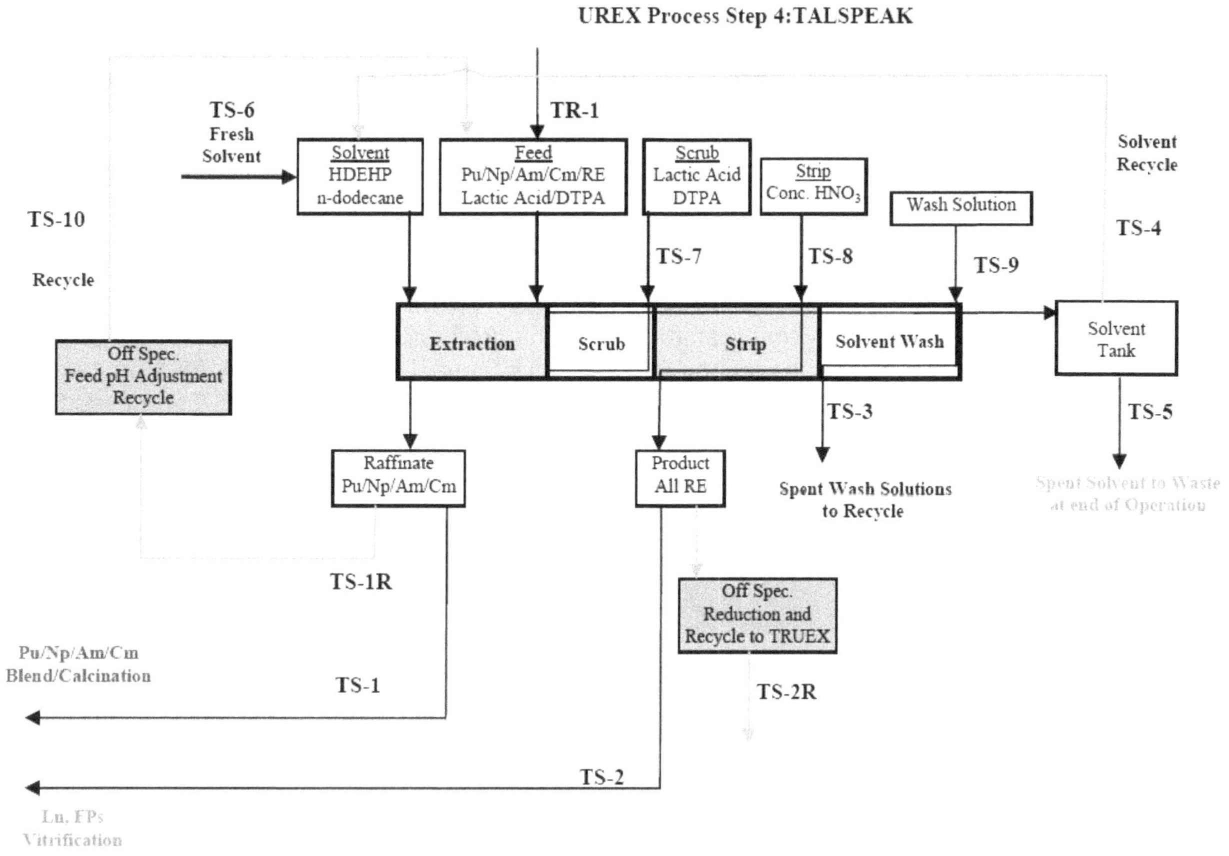

Figure 18: Diagram of UREX+1a Step 4 TALSPEAK to remove lanthanides from TRU

6.1.1. Discussion of the UREX+1a Flowsheet

This section elaborates on the four process steps in the UREX+1a flowsheet and culminates in a description of the products, wastes, and separation efficiencies in the process steps. Major intermediate process stream compositions, recycle streams, and waste streams, and their purity and impurities are presented and discussed insofar as they were available as of February 2007 and are in the publicly available literature.

Although each of the four solvent extraction separation steps has been separately studied experimentally and some have reached advanced stages of development, very little data on the efficiency and operability of the integrated separations are available. Except for the UREX separation step for uranium and technetium, which is essentially a modified PUREX process, no large-scale operating experience is available with the various steps of the UREX processes.

6.1.1.1. Head End

Table 18 describes the key characteristics of a PWR fuel assembly that would constitute the feed to a reprocessing plant [Croff, 1978].

Table 18: Typical PWR Assembly Composition

Fuel Assembly Component		Mass, kg
Fuel material		
Uranium (expressed as elemental U)		461.4
Uranium (expressed as the dioxide)		523.4
Hardware		
Zircaloy-4 (cladding, guide tubes)		108.4
Stainless steel 304 (end fittings)		17.1
Stainless steel 302 (plenum springs)		21.9
Inconel-718 (grid spacers)		5.9
Nicrobraze 50 (brazing alloy)		1.2
	HARDWARE TOTAL:	154.5
	FUEL ASSEMBLY TOTAL:	677.9

Note the large amount of hardware that must be disposed of as radioactive waste. In the head-end step of conventional reprocessing of spent LWR fuel, in the head-end step, the spent fuel is removed from the storage area and segmented before it is dissolved in nitric acid in the head-end hot cell. The assembly may be broken down into individual fuel elements or sheared as a whole. Segmentation is typically done with a remotely operated shear that cuts the spent fuel elements or assemblies into pieces 1 to 2 inches long. This permits ready access of the nitric acid dissolvent to the oxide fuel pellets in the cladding.

During fuel segmentation and dissolution, gases or volatile fission products trapped in the fuel or present in the plenum space at the ends of the fuel elements are released into the hot cell off-gas system. For waste management, the most important off-gas species are ^{129}I, ^{85}Kr, ^{3}H, and

$^{14}CO_2$. The Zircaloy cladding hulls will contain an appreciable fraction of the tritium (as much as 41 percent) as zirconium hydride. Some volatile RuO_4 may also be present in the off-gas. Although the radioactivity of ruthenium isotopes in aged spent fuel is low (1.851×10^{-2} Ci/MTIHM after 25 years of decay), the total mass of ruthenium is not negligible (8.691×10^2 g/MTIHM). For this reason, it deserves attention because it may interfere with recovering the important off-gas species noted above. Because of the ease of reduction of the volatile RuO_4, it may be removed from the off-gas by trapping on steel wool filters which become a waste.

With the exception of iodine trapping processes, many of the candidate processes potentially applicable to U.S. reprocessing plants for trapping the other volatile fission products are in an early stage of engineering development and demonstration, although some of the technology such as cryogenic processes for recovering noble gases is well known in other applications. Iodine trapping methods include scrubbing the dissolver off-gas with a KOH solution, highly concentrated nitric acid, or mercuric nitrate solution, or trapping on solid sorbents, principally those containing silver with which iodine reacts to form highly insoluble AgI or $AgIO_3$. Sorption on charcoal has been used, but charcoal has significant drawbacks primarily because of its flammability. Only the very long-lived ^{129}I iodine isotope ($t_{1/2} = 1.57 \times 10^7$ yr) is of consequence in spent fuel reprocessing because the other iodine isotopes are either very short-lived (^{131}I: $t_{1/2} = 8.02$ days) or stable. A fraction of the iodine may remain in the dissolver solids as AgI and PdI_2. This residue may be put into solution and subsequently into the off-gas by the addition of KIO_3 to the dissolver, but this would require another process step.

The only krypton isotope of radiological importance in SNF reprocessing is ^{85}Kr ($t_{1/2} = 10.72$ yr). Krypton removal has been studied using cryogenic distillation, sorption on zeolites and charcoal, and selective sorption in various liquids such as dichlorodifluoromethane (a refrigerant now out of favor because of its effect on the ozone layer). Diffusion through permselective membranes such as silicone rubber is also a candidate for krypton separation. Xenon, which has negligible radioactivity in long-cooled fuel, has about 19 times the volume of krypton in the off-gas after 25 years of decay. Both of these gases are chemically inert, and their physical properties are the basis of their separation from other gases. However, it is possible to separate krypton from xenon and thus reduce the volume of the radioactive rare gas stored.

Tritium ($t_{1/2} = 12.26$ yr) is a rare isotope in the natural environment. About two-thirds of the tritium produced in LWR fuels is from ternary fission and one-third from neutron activation of lithium. During aqueous reprocessing of spent LWR fuel, any tritium that has not reacted with oxygen in the fuel or escaped as gas in the head-end step will react with water in the dissolver and produce tritiated water, HTO. A promising method for controlling tritium during fuel reprocessing is voloxidation [Goode, 1973a], which Section 6.1.2.1 describes in more detail. In voloxidation, the tritium is vaporized from the spent fuel by heating in air or oxygen before spent fuel dissolution in acid. The HTO thus formed may then be trapped in a dessicant such as silica gel or a zeolite. If tritium removal and containment are required for plant licensing, then voloxidation may be the removal method of choice. If tritium is not removed before acid dissolution of the fuel, then it exchanges with hydrogen in the acid in the dissolver solution to produce tritiated water whose disposal path would be through evaporation. This may not be an acceptable approach. In any case, the relatively short half-life of tritium means that after 100 years, it will have decayed to a very low level of radioactivity.

Spent fuel contains ^{14}C ($t_{1/2} = 5.73 \times 10^3$ yr), which is primarily produced from the ^{14}N (n,p)^{14}C reaction with the nitrogen that is typically present in the fuel at a level of 10–60 parts per million. ^{14}C is produced at a rate of about 10–20 Ci/GWe/yr of reactor fuel irradiation [Choppin, 1987]. Its removal is a straightforward operation in principle because the carbon will be present as $^{14}CO_2$, which is readily sorbed in a large number of sorbents such as KOH, CaO, and molecular sieves (zeolites).

The above discussion shows that, because of their short half–lives, neither krypton nor tritium is a long-term hazard. Storage for 100 years would suffice to remove them from further concern. On the other hand, if capture and storage are imposed requirements for iodine and carbon, they will remain as long-term concerns. At present, there are no generally accepted chemical forms or methods for their permanent disposal.

6.1.1.2.   Central Unit Operations

6.1.1.2.1.   UREX

In this report, the first step in the UREX+1a process is simply called UREX. In the UREX step of the DOE UREX+1a process, the uranium and technetium in solution[23] are separated by solvent extraction with TBP, typically as a 30-percent by volume solution in n-dodecane, from the other actinides, the lanthanides, and the fission products. Technetium extracts along with zirconium as a complex species.[24] The addition of the reducing agent AHA in the process prevents the extraction of plutonium by reducing it to in-extractable Pu(III). After being stripped into an aqueous stream with nitric acid, the uranium is converted to oxide for storage and subsequent use or disposal. If the AHA is omitted in UREX, the process becomes essentially the PUREX process because the uranium and plutonium would be co-extracted in purified form and can be readily separated.

The use of pulse columns for solvent extraction leads to process simplicity and reliability. However, centrifugal contactors can process a given amount of spent fuel faster and in a much smaller space at the cost of increased complexity and somewhat decreased reliability. Specifically, centrifugal contactors cannot tolerate "crud" accumulation because it tends to block overflow orifices. A small amount of solid noble metals has been observed to precipitate slowly from the dissolver solution,[25] and this could pose problems in a centrifugal contactor.
The volume of solid waste produced is related to the type of reagents used in reprocessing. For example, although the PUREX process uses TBP, neither the TBP nor its degradation products can be converted entirely to gaseous products because of the presence of the phosphorus atom in the molecule. This leads to a nonvolatile solid waste.

The UREX+1a process removes the technetium from the acidic uranium product stream using an organic anion exchange resin (technetium is present as the TcO_4^- anion). The TcO_4^- anion is stripped from the resin and precipitated as finely divided metal by use of an alkaline solution of

---

[23] The pertechnetate anion, TcO_4^-, is thought to form an extractable complex species with zirconium which upon extraction releases the pertechnetate ion, which then forms a complex species with the uranyl ion (UO_2^{2+}) and remains largely, but not entirely, within the uranium stream.

[24] Notwithstanding the experience with incomplete extraction of technetium observed by others, in THORP it was found that essentially all of the technetium extracted with the uranium. Changes in process chemistry made it possible to strip technetium selectively from the uranium by using high-acidity in a technetium contactor.

[25] Although this delayed precipitation of noble metals has been observed in early work at ORNL, it has not been observed in THORP operations, even though it was specifically sought.

sodium borohydride or by reduction to metal in a furnace. After multiple uses and stripping to remove residual technetium, the anion exchange resin is carbonized, packaged, and shipped off site for disposal. The technetium metal may be converted to a final waste form by combining it with the washed and compacted cladding hulls from the head-end dissolution step. Alternatively, it could be combined with the noble metal dissolver solids and disposed of with that waste.

6.1.1.2.2. CCD-PEG

The raffinate from UREX contains the actinides plutonium, neptunium, americium, and curium, as well as the lanthanides, ^{137}Cs, ^{90}Sr, and other fission products. The raffinate from UREX becomes the feed to process step 2, the CCD-PEG process [CCD-PEG, 2003, 2006], where the cesium and strontium are separated from the actinides, lanthanides, and fission products using a CCD-PEG solvent as extractant. The CCD-PEG process is most efficient when the feed is less than or equal to 1 M nitric acid so it can be used directly on the low-acidity UREX process step raffinate. The separated cesium and strontium may be solidified in several ways, including as stable aluminosilicate waste in a steam reforming process using an incorporated clay such as kaolin to reduce the solubility of the cesium and strontium.

6.1.1.2.3. TRUEX

The raffinate from process step 2 becomes the feed to process step 3, the TRUEX process [TRUEX, 1998], where the TRU actinide and lanthanide elements are extracted from the remaining fission products using a mixture of TBP and carboxylmethylphosphine oxide in n-dodecane extractant. The actinides and lanthanides are stripped from the extractant with lactic acid. The strip solution becomes the feed to the next and final UREX+1a process step.

6.1.1.2.4. TALSPEAK

The strip solution from the TRUEX process containing the actinides and lanthanides becomes the feed to process step 4, the TALSPEAK process, where, after feed adjustment, the lanthanides are extracted from the actinides [TALSPEAK, 1964, 1999]. The TALSPEAK process performs the difficult separation of actinides and lanthanides, whose chemistries are very similar. This solvent extraction separation process is carried out using Bis-(2-Ethylhexyl) Phosphoric acid (HDEHP) in n-dodecane as extractant, with lactic acid and diethylenetriaminepentaacetic acid (DTPA) as complexants, and concentrated nitric acid as a stripping agent. Very careful control of pH at about pH 3 and careful control of organic-to-aqueous process stream phase ratios are required to effect the desired separation.

The TALSPEAK process relies on the difference in the strengths of the respective complexes formed by the lanthanides and the actinides with DTPA to achieve their separation. The DTPA complexes are not extracted. Because a much smaller fraction of the lanthanides are complexed, HDEHP extracts a larger fraction of them.

The following chemical reaction can be used to illustrate the strong dependence on pH of complex formation with DTPA:

$$M^{n+} + H_5DTPA \rightarrow MDTPA^{(5-n)-} + (5-n) H^+$$

Here M represents the actinide or lanthanide ion, n is the valence of the species involved, and H is hydrogen in the reaction. From this equation, it is apparent that for trivalent ions, there is a square dependence on the hydrogen ion concentration. Thus, if the pH goes from 3 to 4 (i.e., if it changes by a factor of 10), the equilibrium shifts by a factor of 100 to the left, assuming that all else stays the same. This helps explain the exceptional sensitivity of the TALSPEAK process to pH.

6.1.1.2.5.   Products and Wastes

The TRU elements are in the raffinate from the TALSPEAK extraction cycle. They are to be solidified, possibly in combination with some of the uranium, packaged, and stored until refabrication into fuel for transmutation. The lanthanides and residual fission products are in the strip stream and are solidified, packaged, and stored until the time of final disposal.

The lanthanides (also called rare earths) are the radionuclides selected by both the UREX processes and the French GANEX (see Section 6.3.4.1) process for separation from the actinides because of their interference with efficient recycle and reuse of the actinides.

Cesium and strontium wastes are to be put into a stable chemical form and stored for their eventual decay to levels acceptable for near-surface disposal. In this scenario, it will be necessary to provide monitored storage space for the cesium and strontium for an extended time.

A small amount of fluoride (about 0.01 M) is used in the dissolution step because after fuel dissolution, the acidity is reduced during feed adjustment to the point that a fluoride ion is needed to prevent hydrolysis (through complexation) of some of the radionuclides. Although not listed in the flowsheets, a fluoride ion appears in the feed and the raffinate streams in all the process steps. The fluoride ion can exacerbate corrosion, especially in equipment like the dissolver and the waste vitrifier.

6.1.2. Process Assumptions for Modeling the UREX+1a Flowsheet

To calculate the distribution of radionuclides among the waste and product streams, it is necessary to make some assumptions about separation factors achieved in the process steps. There has been considerable experience in reprocessing, and some separation factors are known for common processes like PUREX. The major spent fuel reprocessors (e.g., France and the United Kingdom) consider the separation factors to be proprietary information. However, the *Code of Federal Regulations* or consensus product specifications do identify certain limits on the concentration of radioisotopes in wastes. In the absence of data on separation factors, these limits may be used as criteria that must be met, and thus as specifications for the wastes. Additionally, for some of the less common UREX+1a process steps (e.g., CCD-PEG, TRUEX, and TALSPEAK), publications discussed earlier contain information from laboratory experiments or on limited plant experience that may be used to derive separation factors. All of these sources of information, along with information from burnup calculations made with ORIGEN2 [Croff, 1980] and the judgment of the authors, were used to obtain the process assumptions for modeling the UREX+1a flowsheet contained in Appendix E.

The following sections discuss the most important product, effluent, and waste streams that would be produced by a reprocessing plant using a UREX+1a flowsheet.

6.1.2.1.   Off-Gas Effluent Stream

All plant operating areas have off-gas systems that capture the gases and vapors leaving the area and treat them before they are vented to the atmosphere. In general, air flows from areas of low radioactivity to areas of higher radioactivity to reduce contamination. Each vented radionuclide has a different biological effect on the human body, and this must be considered when deciding what action to take for that radionuclide. In general, the radionuclides in the off-gas must be retained at least to the level of retention required by the regulations. These limits and the technologies proposed to meet them and to retain the radionuclides for storage and disposal have been discussed [ANL 1983; DOE 1986; Goode 1973 a, b; IAEA 1980, 1987, 2004; Wigeland 2007].

The most important reprocessing off-gas streams are those from the spent fuel shear and the dissolver. These streams contain the bulk of the radioactive gases and vapors (tritium, krypton, iodine, carbon dioxide, ruthenium, particulates, and aerosols), as well as hazardous chemical species (nitrogen oxides). Other important off-gas streams are those from the fission product and lanthanide waste calcination (if used) and vitrification steps, which this paper does not examine. Numerous specific technologies can remove these species from off-gas streams.

- Tritium [IAEA, 2004; DOE, 1986]: To be effective, recovery of tritium must occur before the spent fuel encounters substantial amounts of water, such as the dissolver solution, to prevent isotopic dilution of the tritium with large amounts of ^1H in water. As a consequence, tritium removal and recovery occur immediately after the spent fuel is chopped (sheared) into segments using the voloxidation (volume oxidation) process. This process depends on the oxidation of the UO_2 spent fuel matrix to lower density U_3O_8 to break down the fuel matrix and release trapped gases from it. (Voloxidation is unlikely to be effective with thorium-based fuels because thorium does not have a higher valence state to which it can be oxidized.) Voloxidation is implemented by heating the spent fuel segments to 450 to 500 °C or possibly higher for several hours in a rotary kiln. The tritium in the evolved gas is passed through a catalytic converter to yield tritiated water, which is then removed from the off-gas by solid dessicants. Essentially all of the tritium is released from the spent fuel (but not necessarily from the Zircaloy hulls), and much smaller fractions of other volatile species are released as well. If dehumidified oxygen is used in the kiln, then the recovered tritium will be very concentrated. To the extent that humidity is introduced, the tritium will be diluted and the volume of the tritium waste form increased.

  Development of voloxidation had largely ceased for about two decades at the end of the 1970s. However, DOE is now supporting work in the United States and South Korea to further develop voloxidation and South Korea has a collaborative effort with Canada to develop the DUPIC process (see Section 3.1.2.7.) that supplements the DOE effort. The goal is to maintain the high release rates for tritium while increasing release rates of other volatile species. Variations being examined include use of temperatures up to about 800 °C; use of some combination of air, ozone, and steam to oxidize the fuel; and cycling between oxidizing conditions and reducing conditions imposed by hydrogen gas in the voloxidizer.

  Important open technical issues concerning voloxidation are the extent to which tritium is evolved from zirconium tritide formed in the Zircaloy cladding during voloxidation and the extent to which other volatile species will be evolved from the fuel matrix. Also, the effectiveness of voloxidation on fuels containing high concentrations of TRU elements such as those that might be used in a transmutation reactor is largely unknown.

- Iodine, Ruthenium, Aerosols, Particulates, and Nitrogen Oxides [IAEA, 1987; DOE, 1986]: After voloxidation, the spent fuel segments are loaded into a dissolver containing concentrated nitric acid which results in evolution of the volatile radioactive and hazardous species other than tritium from the dissolver vessel into the off-gas. The next step in treating the off-gas is to remove aerosols and particulates, nitrogen oxides, ruthenium, ^{129}I, and then more nitrogen oxides in that order. This is accomplished by passing the off-gas through a water scrubber and de-entrainer to remove most of the nitrogen oxides as well as aerosols and some particulates. The off-gas is then heated above its dewpoint and passed through a silica gel bed to absorb ruthenium[26] and a HEPA filter for additional particulate removal. The off-gas stream is passed through sequential beds of silver zeolite to remove iodine. Although iodine decontamination factors of greater than 99.5 percent have been achieved at La Hague and THORP using caustic scrubbing, it has not yet been shown that large reprocessing plants in the United States will actually be able to achieve this performance using the proposed processes. Finally, the off-gas is further heated, mixed with ammonia injected into the waste stream, and passed through a zeolite bed which decomposes the residual nitrogen oxides and ammonia to nitrogen and water.

  Alternative iodine removal technologies have been developed and demonstrated or used in small-scale plants. The advantages and disadvantages of various iodine removal processes are discussed in [DOE, 1986] and [IAEA, 1987].

- ^{14}C [IAEA, 2004; DOE, 1986]: The off-gas from the iodine removal step flows through two molecular sieve beds connected in series for water removal followed by two zeolite beds connected in series for CO_2 removal. Water is removed from the sieves by reducing the pressure. A similar approach is used for the zeolite beds containing the $^{14}CO_2$. The resulting concentrated carbon dioxide stream is routed to a scrubber where it bubbles through a saturated solution of $CaOH_2$ to form insoluble calcium carbonate containing the ^{14}C. The calcium carbonate is recovered using a vacuum filter, dried, and stabilized in drums.

- ^{85}K [IAEA, 1980; DOE, 1986]: The off-gas feed stream to the krypton recovery system consists primarily of air with small amounts of water, nitrogen oxides, radioactive krypton, and stable xenon. The oxygen in the air is removed by reacting it with hydrogen in a catalytic recombiner. The gas is refrigerated to condense some additional water and then passed through silica gel for final water removal. The off-gas then enters a cryogenic absorption, stripping, distillation, and recovery process. Liquid nitrogen is the primary working fluid to enrich the krypton concentration relative to that of xenon from about 7 percent at the outset to about 80 percent in the product. The krypton-xenon product is then packaged for disposal.

---

[26] The significant radioactive isotope of ruthenium (mass number = 106, half-life = 1 yr) is only relevant in fuels aged less than about 10 years before reprocessing, which may not be the case in the United States for many years. However, nonradioactive ruthenium removal may still be needed to prevent clogging of the off-gas system.

- <u>Particulates</u>: The final off-gas treatment step is additional HEPA filtration to remove the remaining particulates and aerosols.

6.1.2.2. Technetium Stream

Historically, conventional wisdom held that technetium would not extract quantitatively with the uranium in the first process step. However, experience at THORP (see Sect. 3.1.3.7) indicates that technetium does extract quantitatively and could be readily recovered from the uranium stream. Also, as much as 15 percent of it may become part of a noble metal (e.g., palladium, ruthenium, rhodium, platinum) sludge in the spent fuel dissolver, in which case that portion could be managed by combining it with the cladding hulls as shown in the UREX+1a flowsheet or separately by means to be determined. Addressing these issues requires more definitive experimental information on the form and distribution of technetium in UREX that requires results from an integrated engineering flowsheet demonstration and optimization.

6.1.2.3. Uranium Product Stream

The uranium product stream contains 2097 MTHM of uranium (as uranyl nitrate) annually from a 2200 MTIHM/yr reprocessing plant. There will need to be a substantial uranyl nitrate denitration system to convert the liquid uranyl nitrate to solid uranium oxide. Denitration will produce nitrogen oxides, which must be recovered to prevent escape of toxic NO_x gases to the atmosphere. There is also the option of making nitric acid from the nitrogen oxides.

6.1.2.4. Solvent Waste Streams

There will be enough radioactivity in the solvent waste streams to require care in their disposal. As noted earlier, each UREX+1a process step has a different solvent, and each probably requires a different waste cleanup system. As the solvents need to be replaced, solvent waste streams will be produced. Incineration may possibly be an acceptable means for treatment of most of them because almost all the solvents are organic compounds. However, UREX and TALSPEAK process steps contain solvents (i.e., TBP and HDEHP) that cannot be completely oxidized to gaseous compounds.

6.1.2.5. Fission Product Stream

The fission product waste stream, as the term is defined in this paper, contains all the fission products except cesium, strontium, technetium, iodine, krypton, tritium, and carbon. These wastes are primarily the lanthanides and are the remaining wastes to be vitrified, packaged, stored, and ultimately sent to a deep geologic repository.

6.1.2.6. Cesium/Strontium Stream

^{137}Cs and ^{90}Sr pose a special and significant waste management problem. Together, they are a major medium-term heat producer (see Figure B1 in Appendix B), because they account for more heat and more radioactivity than all the other radionuclides for several decades. ^{137}Cs is a source of penetrating radiation[27] and merits special attention. It is apparent that the cesium and strontium constitute a major waste management problem. The cesium/strontium is to be fixed in a chemically stable waste form, packaged, stored for about 300 years to allow it to decay to less-than-Class C concentrations, and then disposed of in place.

6.1.2.7. Actinide Stream

The actinides are the principal useful product of the reprocessing plant, as well as being a principal heat source (see Appendix F and the graph in Appendix B). About 27.7 MTHM per year of actinides from a 2200 MTIHM/yr reprocessing plant (exclusive of any uranium that might be added) will need to be packaged, stored, and ultimately sent to a reactor for transmutation to fission products, which themselves will, after reprocessing, be added to the fission products already produced in the original irradiation that produced the spent fuel.

6.1.3. Quantitative Analysis of UREX+1a Waste and Product Stream Characteristics

The purpose of this section is to provide the results of an illustrative calculation of the radioactive and physical properties of the waste and product streams from the UREX+1a flowsheet. The purpose of such calculations is to approximate the characteristics of typical UREX+1a wastes as a basis for evaluating the work necessary to develop an appropriate regulatory framework for recycle facilities. Such calculations are based on a large number of assumptions concerning, for example, the age and burnup of the SNF fed to the process; separation factors for key radionuclides for each step in the process; and the chemical form, stabilization matrix, loading, and density of the final product or waste forms.

Figure 19 [Kouts 2007] shows the burnup distribution of the spent LWR fuel in 1999 as a function of age. As is evident, the age and burnup cover a wide range. Adjusting the age distribution for time elapsed since 1999 leads to an average age of about 25 years for SNF currently in storage. The reprocessing of SNF would slow or reverse the trend of increasing SNF age depending on whether SNF were to be reprocessed at a greater rate than it is being produced. However, the likely initiation of reprocessing is at least a decade away, which will make the average feedstock commensurately older. Additionally, with SNF being produced at a rate of 2100 MTIHM per year, it would take the equivalent of three large (about 800 MTIHM per year throughput) SNF reprocessing plants just to stabilize the aging of the SNF inventory. Achieving this throughput appears to be some distance in the future because DOE has stated that the throughput of the consolidated fuel treatment center (CFTC) should be able to be increased to approximately 2,000 to 3,000 MTIHM per year to support commercial operation [DOE 2006a]. In a notice requesting expressions of interest in the CFTC [DOE 2006b], DOE implies that the initial throughput will have a value below this range.

---

[27] Although the ^{137}Cs itself is not an important source of radioactivity (beta rays of less than 40 kiloelectronvolts), 92 percent of its decays to ^{137m}Ba which decays with a half-life of 2.55 minutes; 90 percent of the ^{137m}Ba decays to yield a 0.662 MeV gamma ray, which is the source of penetrating radiation.

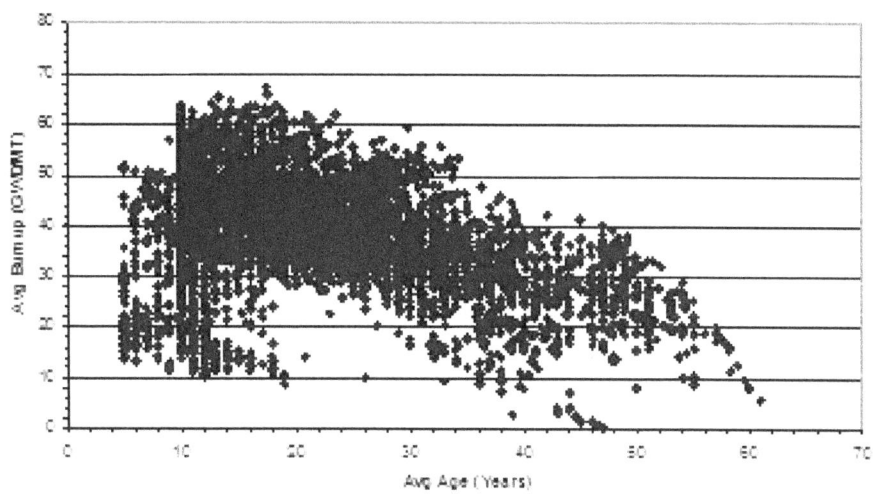

Figure 19: Distribution of U.S. spent nuclear fuel in 1999 as a function of age

The average burnup of stored LWR SNF at the end of 1998 was 30.4 GWd/MTIHM; at the end of 2002 (the latest report), this had increased to 33.6 GWd/MTIHM [EIA, 2004]. The trend of increasing burnup seems likely to continue as utilities seek to get more energy out of their fuel, although high uranium or enrichment costs could mitigate the trend.

On balance, a burnup of 33 GWd/MTIHM has been selected because this value is close to the current average burnup. Additionally, this assumption was efficient and facilitated verification of calculated results by allowing the use of existing PWR models for which published results were available. Given the speculative nature of assumptions concerning when reprocessing might ensue and the rate at which capacity will be built, an SNF age of 25 years was selected.

To calculate the waste compositions and characteristics, it was necessary to use values for separation factors of the various radionuclides in the process steps, as well as waste form densities and loadings. In most cases, reliable values for separation factors from plant operating data are not available. Plant operators usually consider these to be proprietary, although some data from early reprocessing have been published. There are also some data reported in the early literature and data from laboratory experiments using fully irradiated fuel for processes like CCD-PEG, TRUEX, and TALSPEAK. However, the entire UREX+1a flowsheet has not been demonstrated on SNF at a scale that provides a reliable foundation for assumptions concerning separation factors. Regarding parameters concerning waste form characteristics, in many cases fundamental decisions (e.g., which waste form will be used) have not been made. Based on evaluation of the results of UREX+1a experiments performed to date and the historical and current literature by independent experts concerning waste forms, the values and assumptions in Appendix F were assembled. These values and assumptions were used to calculate the waste stream compositions amounts using ORIGEN2. Table 19 gives the results of these calculations.

During review, many comments indicated the desire for additional detail on the composition of internal plant streams that might be important to safety. While the quest for such data is understandable, the UREX+1a flowsheet and waste treatment and disposal requirements are not yet sufficiently mature to allow the determination of such compositions. Additionally, such information is likely to be deemed sensitive and/or proprietary and could not be included in a public document such as this. Consequently, this paper does not include information at this level of detail.

Table 19: Compositions and Amounts of Waste Streams

OUTPUT	AMOUNT PER MTIHM FROM ORIGEN2				FINAL OUTPUT STREAM CHARACTERISTICS				
	Grams	Curies @ 25 yr	TRU α curies @ 25 yr	Watts @ 25 yr	Density (g/cc)	Grams nuclide/ Gram Waste	Waste Volume (L/MTHM)	TRU nCi/g	Classification/ Relation to Class C[j]
Volatiles Release									
T	0	0	0	0					
C	0.013	0.01	0	0					
Kr	0.7	277	0	2.35					
I	0.9	0	0	7.00E-08					
Volatiles in Waste									
T[a]	0.021	201	0	0.01	2.2	0.079	0	0	Class B/4e+8x[k]
C[b]	0.133	0.592	0	0	1.6	0	1.81	0	GTCC/41x
Kr[c]	4	1570	0	2.35	0.005	0.011	77.4	0	Class A/Not Listed
I[d]	177	0.031	0	0	2.1	0.0414	2.4	0	GTCC/163x
Cladding + Tc[e]	296000	1020	3.33	4.97	6.8	1	43.5	11000	GTCC/220x
U Product[f]	953000	8.21	0.01	0.088	3.5	1	272	5	Class A/0.05x[n]
TRU Product[g]	12600	44400	6654	222	10.8	1	1.17	5.30E+08	HLW/5e+6x
Cs/Sr Waste[h]	5150	154000	0	328	1	0.27	19.1	0	HLW/1570x[o]
Fission Product Waste[i]	19700	42300	1.41	235	2.65	0.38	19.6	27000	HLW/270x
Spent Nuclear Fuel	1.45e+6[l]	242600	6660	789	5	1	403[m]	4600000	HLW/46000x

From previous table:
[a]Tritiated water in polymer-impregnated cement
[b]Calcium carbonate in cement
[c]Compressed gas in cylinder
[d]Grouted silver zeolite
[e]Homogeneous alloy of structural material, dissolver solids, Tc, and some undissolved fuel
[f]Uranium oxide having concentrations of other radionuclides suitably low to allow re-enrichment
[g]Oxides of the various elements sintered to 95% of theoretical density
[h]Glass-bonded aluminosilicate made by steam reforming
[i]Vitrified into borosilicate glass logs
[j]Apparent waste classification/ratio of sum-of-the fractions for material to Class C limits if material is declared to be waste
[k]Assumes no dilution of tritium with hydrogen isotopes other than those produced in the fuel matrix
[l]Includes oxygen associated with fuel matrix
[m]Volume of a single, intact PWR SNF assembly (21.4 cm x 21.4 cm x 406 cm) normalized to 1.0 MTIHM. Volume of an intact assembly in a canister would be 635 liters per metric ton of heavy metal (L/MTIHM).
[n]Assumes that recycle uranium specifications are met for TRU and that Tc concentrations are typical of THORP experience (0.03 ppmw). TRU element concentration can increase about 2-fold before Class C levels are reached and about 20-fold before the uranium would be GTCC.
[o]Waste is HLW because it is derived from the first cycle raffinate unless DOE determines it is waste incidental to reprocessing (WIR).

The following sections discuss waste-specific aspects of Table 19. One generic aspect of Table 19 concerns the decision of which materials would be HLW if they were declared to be waste. The following is the current definition of HLW [NWPA, 1996]:

> The term "high-level radioactive waste" means—
> (1)   the highly radioactive material resulting from the reprocessing of spent nuclear fuel, including liquid waste produced directly in reprocessing and any solid material derived from such liquid waste that contains fission products in sufficient concentrations; and
> (2)   other highly radioactive material that the Commission, consistent with existing law, determines by rule requires permanent isolation.

Key terms such as "highly radioactive" and "fission products in sufficient concentrations" have not been further elaborated. Additionally, the Commission has not identified any "other highly radioactive material" that requires permanent isolation. Appendix F, "Policy Relating to the Siting of Fuel Reprocessing Plants and Related Waste Management Facilities," to 10 CFR Part 50, "Domestic Licensing of Production and Utilization Facilities," interprets the preceding definition as follows:

> [...] high-level liquid radioactive wastes" means those aqueous wastes resulting from the operation of the first cycle solvent extraction system, or equivalent, and the concentrated wastes from subsequent extraction cycles, or equivalent, in a facility for reprocessing irradiated reactor fuels.

Conventionally, HLW has been taken to include the raffinate from the first cycle of solvent extraction in a PUREX facility. This would include essentially all of the nonvolatile fission products, neptunium, americium, and curium, plus a small fraction of the uranium and plutonium. HLW would not include cladding and other fuel assembly structural materials or volatile species because these are removed before the first solvent extraction cycle. Additionally, HLW would not include organic wastes (e.g., spent solvents.) Historical designs for PUREX reprocessing plants have typically found it convenient to concentrate some of the more active waste streams from parts of the reprocessing plant other than the first solvent extraction cycle and combine them with the aqueous waste from the first cycle of solvent extraction and manage them all as HLW.

The existing definition of HLW was not conceived with the UREX flowsheets in mind. In particular, part of the UREX+1a product (the TRU actinide elements) is initially in the aqueous waste from the first cycle of solvent extraction and becomes a separate product stream only after going through three subsequent solvent extraction processes. Additionally, separation of cesium/strontium from the aqueous waste from the first cycle of solvent extraction raises the question of whether the separated cesium/strontium is HLW. Based on historical and recent (e.g., concerning "waste incidental to reprocessing") interpretations of what constitutes HLW, this paper assumes that any material contained in the aqueous stream from the first cycle of solvent extraction that is declared to be waste would be classified as HLW whether it has been separated from the aqueous stream in subsequent processes or not. The rightmost column of Table 19 reflects this assumption.

6.1.3.1. Volatiles in Waste

Although waste forms for tritium, $^{14}CO_2$, and ^{85}Kr are shown here, these radionuclides have not been sequestered previously because no standards have been in place that specifies that they be recovered and how they should be treated and subsequently disposed. Consequently, these waste forms are the authors' judgment of what might constitute credible waste forms.

Because ^{129}I concentrates in the thyroid gland where, in sufficient amount, it may cause serious damage, especially in children, its sequestration has been required from the beginning of reprocessing. Care is required to ensure its complete release into the off-gas during spent fuel dissolution [CEA, 2007]. It is an especially troublesome radionuclide to dispose of as waste because it has few highly stable chemical compounds. This study chose fixation of the iodine on silver zeolite sorbent because the system is inorganic and therefore less subject to radiation damage than organic materials, AgI is insoluble under most conditions likely to be found in the environment, and AgI is stable to relatively high temperatures (it decomposes at its melting point of 552 °C).

The following information elaborates on issues related to the classification of waste forms containing volatile radionuclides:

- Tritium: Tritium is recovered by voloxidation before encountering the first aqueous solutions in the reprocessing plant. It is assumed to be diluted by only the very small amounts of ^1H and deuterium produced by nuclear reactions in the fuel matrix. The possibility of dilution by water in air used to oxidize the fuel in the voloxidation step was not considered because the amount of humidity and air are design specific. These assumptions lead to a very high tritium concentration in a very small volume of waste.

- ^{14}C: Dilution with stable carbon isotopes in the fuel matrix and some natural carbon in the plant off-gas were considered, but the very small volume of the waste results in ^{14}C concentrations exceeding Class C limits.

- ^{85}Kr: ^{85}Kr is not listed in the tables in 10 CFR Part 61, "Licensing Requirements for Land Disposal of Radioactive Waste," so it is Class A by definition. Such classification deserves further evaluation because the half-life of ^{85}Kr is similar to that of tritium, but ^{85}Kr is more difficult to stabilize and has significant penetrating radiation.

- ^{129}I: Again, the relatively small volume of the waste leads to a high concentration of iodine in the waste form and classification as GTCC.

6.1.3.2.   Cladding, Technetium, and Dissolver Solids

The cladding and technetium wastes may also contain the so-called noble metals platinum, palladium, rhodium, ruthenium, and molybdenum that constitute the dissolver solids. These noble metals may or may not be combined with the cladding hulls. If they are not removed from the dissolver with the cladding hulls, then they will be left in the dissolver and may be carried into the UREX process step. Together, they present a potential problem in that, being solids, they may cause hot spots in the dissolver and subsequently in the centrifuge used to clarify the feed to the solvent extraction equipment. If they persist beyond the feed clarification step, they may cause problems in the centrifugal solvent extraction contactors. A particular problem is the potential blocking of the organic overflow weirs. The dissolver solids problem is further exacerbated by the fact that small amounts of solids have been observed to continue precipitating from the dissolver solution for up to 2 weeks, as noted in Section 6.1.1.2.1 (however, see footnote 24.). Further, if carried into the UREX process step, the solids would add to the radiation damage to the solvent. ORNL investigators observed the amount that slowly precipitates to be as much as 10 percent of the amount that remains initially undissolved in the dissolver.

The cladding waste, which is assumed to contain most of the ^{99}Tc and the dissolver solids (which contain a significant fraction of ^{129}I), exceeds the Class C limit by a factor of 220 and is classified as GTCC. The primary contributors to exceeding the Class C limit are TRU elements and ^{99}Tc, both of which exceed the Class C limit by about a factor of 100. ^{94}Nb exceeds Class C limits by about a factor of 10. ^{59}Ni and ^{14}C are close to the Class C limit (0.5 and 0.3, respectively) and might exceed the limit for very high burnup fuels.

6.1.3.3.   Uranium Product

The uranium may follow several different disposition paths. The DOE plans call for beneficial use of the uranium through its combination with the actinide stream for use in the burner reactor or its reenrichment to produce LWR fuel. Some portion of the uranium may not find a beneficial use, in which case it would be converted to an oxide and managed as a waste in much the same way that DOE currently approaches disposition of enrichment plant tails.

Recycled uranium is not as benign as natural uranium for two reasons. First, no separation process is perfect, and the uranium will contain trace amounts of radionuclides such as ^{99}Tc and ^{237}Np. These radionuclides can become concentrated in enrichment facilities and have been troublesome in the gaseous diffusion plants because they tend to deposit on internal surfaces. Such deposits can complicate maintenance activities to the point that gaseous diffusion plant operators have been reluctant to contaminate their plants with recycle uranium or have dedicated certain plants to recycle uranium enrichment. The current trend away from gaseous diffusion and toward gas centrifuge enrichment makes it much more economical to dedicate part of the plant to recycle uranium.

The second difference between natural and recycle uranium is that the latter contains ^{236}U and ^{232}U. The former is an undesirable neutron poison that detracts from the value of the recycle uranium. The latter is present in very small quantities (typically around 1 ppb) but has a relatively short half-life (72 years), and one of its decay products emits a very energetic gamma ray which leads to higher occupational dose rates during fabrication than do those from natural uranium.

Based on the assumptions in Appendix E, the uranium products from reprocessing would be Class A if they were declared to be waste. Class A for the uranium per se is a default classification because uranium is not listed in the classification tables in 10 CFR Part 61. The major contributor to the uranium product being about 5 percent of Class C limits is the trace amount of TRU elements assumed to accompany the uranium. This paper assumes that the TRU elements are removed from recycle uranium to the point that the uranium just meets but does not exceed applicable specifications for recycle. The concentration of TRU elements could likely be further reduced if required.

6.1.3.4. Transuranium Product

The TRU product stream from the TALSPEAK process is destined for transmutation. It produces about two-thirds as much heat as the cesium/strontium waste stream per MTIHM based on 25-year-old SNF and, as a consequence, requires packaging and storage in a way that permits cooling. Additionally, the alpha activity of this material is sufficiently concentrated so that significant upstream (counter to ventilation air currents inside the facility) mobility of the actinides from alpha recoil can be expected and will need to be considered in the design of the off-gas system.

Under the assumption that materials separated from the aqueous raffinate from the first solvent extraction cycle are HLW, if the TRU product were declared to be waste, it would be HLW and, by concentrating the most toxic actinides into a small volume, would exceed Class C limits by a large factor.

6.1.3.5. Cesium/Strontium Waste

^{137}Cs is a difficult fission product to manage. The radioactivity of its short-lived ^{137m}Ba daughter produces an energetic gamma ray and considerable concomitant heat. Consequently, packaging, storing, shielding, and cooling will be significant problems for many decades. In addition, ^{135}Cs which has a long half-life (2.3×10^6 yr) is present in masses comparable to that of ^{137}Cs after 25 years of decay so the cesium waste package may require indefinitely long confinement.

The preceding comment on heat production holds for ^{90}Sr, although its radiation is softer, and there is no other long-lived strontium radionuclide present. The ^{90}Y daughter is quickly in secular equilibrium and decays with a very short half-life to stable ^{90}Zr. Consequently, there may be merit to adding an additional step to separate the strontium from the cesium to reduce the volume of waste held in long-term disposal, although the UREX flowsheets do not do so.

Under the assumption that materials separated from the aqueous raffinate from the first solvent extraction cycle are HLW, the cesium/strontium waste would be HLW unless DOE goes through the process to determine that it is not HLW. If the cesium/strontium were determined not to be HLW, then it would be GTCC waste because the concentrations of ^{90}Sr and ^{137}Cs initially exceed Class C limits by a large factor. Current DOE plans call for this waste to be stored in some type of monitored near-surface engineered storage facility until it decays to Class C levels or lower, at which time the facility would be deemed to be a disposal facility. The combined ^{90}Sr and ^{137}Cs would decay to Class C limits in about 320 years. This disposal approach raises the issue of whether the cesium/strontium waste would be classified when it is produced at the reprocessing plant or after the extended storage period when the storage facility is converted to a disposal facility.

In 10 CFR Part 61, there is no limit for ^{135}Cs, and establishing such a limit might change its classification. However, the draft environmental impact statement for 10 CFR Part 61 [NRC, 1981] stated a limit of 84 mCi/L for ^{135}Cs, which is significantly larger than its concentration of 18 mCi/L in the cesium/strontium waste. An additional complication with the cesium/strontium waste is that cesium isotopes decay to stable barium, which would make the waste a mixed waste under the Resources Conservation and Recovery Act on the basis of the toxicity characteristic of barium, unless standard leach tests show that the waste form releases sufficiently small amounts of barium.

6.1.3.6.   Fission Product Waste

The fission product waste, which in the present discussion does not include the gaseous and volatile fission products or the cesium/strontium fission product waste, is destined for vitrification in borosilicate glass and eventual disposal in a geologic repository. The ultimate mass of fission product waste would be that listed in Table 19 plus the mass of the TRU product that will be fissioned in a transmutation reactor, plus perhaps a few percent of the uranium mass if the uranium were to be reenriched to produce LWR fuel.

The fission product waste is classified as HLW. It exceeds Class C limits by a factor of 270, which is a much smaller factor than that for the TRU product or cesium/strontium waste and is comparable to the cladding plus technetium waste. The residual TRU in this waste is the cause of its exceeding the Class C limits.

6.1.3.7.   Spent Nuclear Fuel Comparison

To provide some context for the preceding discussion, the characteristics of the PWR SNF that produced the foregoing wastes have been included. The following should be noted:

- The parameters in the left portion of the table (mass, radioactivity, and thermal power) are conserved so that the values for the SNF are just the sum of the various wastes and products with minor differences from rounding. As a result of the intense radioactivity and thermal power of the TRU product (americium and curium in particular) and cesium/strontium waste from UREX+1a, the wastes destined for disposal in a deep geologic repository (cladding and fission product waste) are reduced to 18 percent and 30 percent of the amount in SNF respectively. This reduction would not occur for a PUREX process, where the cesium/strontium, americium, and curium remain with the waste destined for deep geologic disposal.

- Assuming that the uranium is reused, the waste volume from UREX+1a would be reduced by about 79 percent as compared to SNF if the relatively voluminous ^{85}Kr is excluded and by 59 percent if it is included. To the extent that uranium is not reused, the volume of reprocessing wastes would be increased, and in the limiting case, the total waste volume would be increased by about 8 percent as compared to SNF. In the case of a conventional PUREX process, the volume of waste destined for deep geologic repository disposal (about 450 L/MTIHM [Vernaz, 2006]) is about the same as the volume of the parent SNF fuel (403 L/MTIHM) per se and less than the volume of an SNF assembly in a canister (635 L/MTIHM). This reduction has been accomplished through careful management of facility operations; use of chemicals that can be degraded to water, nitrogen, and carbon dioxide; and the use of compactors and incinerators. However, to the extent that the uranium product is declared to be waste (up to 272 L/MTIHM) or LLW destined for near-surface disposal (about 200 L/MTHM), total waste volume ranges from 1.45 to 2.3 times that of the SNF depending on which SNF comparison basis is selected.

- The SNF assembly is about 46,000 times Class C limits. This factor is much less than the factor for the TRU product, which reflects the concentration of the most hazardous 10 CFR Part 61 species in the relatively small volume of the TRU product.

6.1.4. Potentially Toxic and Reactive Materials

In general, the Occupational Safety and Health Administration will regulate the nonradiological hazards involved in SNF recycle. The solvents used in the four UREX+1a process steps are commercially available organic compounds and as such require the same handling procedures in a reprocessing plant as those that are required for safe handling of these somewhat toxic chemicals in industrial operations. In ordinary chemical process use, none is extraordinarily toxic or reactive, although all pose some danger to those who handle them. Other chemicals, such as those used in solvent cleanup, are inorganic compounds, and safe industrial practice should be observed. In cases where solvents such as halogenated compounds are used, the toxic halogens may be released by radiolytic decomposition. Thus, although the compounds may be relatively benign in ordinary use, they can become toxic in radiation environments. Nitric acid in a variety of concentrations is used throughout the process steps, and because of its amounts and ubiquity, it is probably the most significant toxic chemical. "Red oil," which is discussed below, presents a significant potential chemical hazard.

6.1.4.1. Red Oil Explosions

Red oil is a substance formed when an organic extractant such as TBP comes in contact with concentrated nitric acid (greater than 10 M) at a temperature above 120 °C. Contributory chemicals can include diluents (e.g., hydrocarbons used to dilute TBP) and/or aqueous phase metal nitrates. Red oil can decompose explosively when its temperature is raised above 130 °C. Three red oil explosions have occurred in the United States (one at the Hanford Site in 1953 and two at the SRS in 1953 and 1975). A red oil explosion also occurred in 1993 at the Tomsk-7 site at Seversk, Russia, and in an evaporator in Canada. Equipment capable of producing red oil includes evaporators and denitrators.

Controls for prevention or mitigation of a red oil explosion are generally controls on temperature, pressure, mass, reactant concentrations, and agitation of tank contents. Maintaining a temperature of less than 60 °C is generally accepted as a means to prevent red oil explosions. Vessel venting serves to keep pressure from destroying the process vessel in the case of an explosion, while also providing the means for evaporative cooling to keep red oil from reaching the runaway temperature. Mass controls utilize decanters, hydrocyclones, and steam stripping to remove organics from feed streams entering process equipment capable of producing red oil. Limiting the total available TBP is another mass control that mitigates the consequence of a red oil explosion by limiting its maximum available explosive energy. Washing the aqueous plutonium and uranium products with diluent to remove entrained TBP is effective in preventing red oil explosions during evaporation of these products. Finally, concentration control can be utilized to keep the nitric acid below 10 M. A U.S. government study [DNFSB, 2003] concluded that none of the above controls should be used alone; rather, they should be used together to provide effective defense in depth for prevention of a red oil explosion. The operator of French reprocessing plants (AREVA) recently stated [ACNW&M, 2007] that red oil has not been observed in its plants.

At present, there is no information about the likelihood of forming red oil in UREX+1a processes, although the first step that uses conventional TBP extraction may be expected to pose the same red oil risks as have been observed in the past.

6.1.4.2.   Ion Exchange Resin Explosions

Nine documented incidents of fire, explosion, and/or vessel rupture in anion exchange vessels at the SRS have been characterized as "resin explosions" [DNFSB, 2001]. They have occurred under various conditions of temperature and nitric acid concentration. All of the systems involved were exchanging ions containing plutonium, neptunium, curium, or uranium.

Conditions identified as contributing to a possible resin explosion are listed below:

- exposure of resin to greater than 9 molar nitric acid

- exposure of resin to high temperature

- allowing resin to dry

- exposure of resin to strong oxidants other than nitric acid, such as permanganate or chromate ions

- exposure of resin to high radiation doses

- allowing resin to remain in a stagnant, nonflow condition while loaded with exchanged metal and/or in contact with process concentrations of nitric acid

- exposure of resin to strong reducing agents, such as hydrazine

- exposure of resin to catalytic metals such as iron, copper, or chromium

By avoiding the above conditions, it was possible to prevent further explosions, but great care must be taken to prevent these explosions in the future, especially if attempting separations involving concentrated americium and curium.

6.2. Pyroprocessing

There are many manifestations of pyroprocessing in the nuclear industry [NEA, 2004], several of which are directed at spent fuel recycle. As applied to SNF reprocessing, pyroprocessing involves the use of molten salts and metals in an electrochemical cell to separate the SNF constituents. Pyroprocessing has been in general use for many years for purification of nuclear materials, including plutonium. It involves anodization (oxidation) of a metal feed material into a molten salt electrolyte and then reduction at a cathode to yield a more (highly) purified form.

Pyroprocesses are not currently in significant use worldwide, but they have been the subject of much R&D. ANL has studied and developed electrometallurgical spent fuel reprocessing for many years, and a demonstration is still underway at the DOE INL facility using Experimental Breeder Reactor II (EBR-II) spent fuel. An NAS committee evaluating the flowsheet (see Figure 20) for this demonstration found no technical barriers to the use of electrometallurgical technology to process the remainder of the EBR-II fuel [NAS, 2000].

The feed to pyroprocessing was originally intended to be metallic spent fuel, and the process lends itself best to reprocessing this type of fuel. As a consequence, the current DOE plans call for pyroprocessing to be used to reprocess metallic or possibly nitride SNF containing the TRU actinide elements after irradiation in a fast-spectrum transmutation reactor. However, oxide fuels such as those from LWRs can be pyroprocessed by first converting them to metal through a head-end step that reduces the oxide to metal. This reduction is best accomplished using finely divided oxide, which can be prepared using voloxidation (see Section 6.1.2.1) to pulverize the oxide fuel. Process modifications are possible that separate uranium, plutonium, and other actinides from the remainder of the radionuclides. Figure 21 [ANL, 2002] represents the pyroprocessing flowsheet for oxide SNF under development by ANL and other organizations such as Korean Atomic Energy Research Institute (KAERI). The following are the major steps in this flowsheet:

- Oxide SNF is chopped into segments and voloxidized (not shown).

- Most of the oxides in the SNF are reduced to the metal. This is accomplished by chemically reducing the SNF oxides using molten lithium (existing technology) or by electrolytic reduction in molten lithium chloride (technology under development by ANL). GE-Hitachi plans to demonstrate electrolytic reduction of uranium dioxide in the near future. Bench-scale tests have shown that about 99.7 percent of the SNF is reduced.

- The metal from oxide reduction or metallic SNF, including the cladding in either case, becomes the anode in an electrorefiner. The electrorefiner is essentially a crucible containing a molten electrolyte salt (a mixture of LiCl and KCl) atop a layer of cadmium metal. The anode and two cathodes operating at different voltages are inserted into the molten salt. After operating for about 12 hours, the electrorefiner contains the following:

  – The anode contains elements that are stable as metals under the conditions in the electrorefiner (e.g., zirconium, technetium, iron, molybdenum).

  – One cathode contains most of the uranium as metal.

  – The other cathode contains some of the uranium and rare earth fission products plus essentially all of the TRU elements as metal.

101

– The molten salt contains most of the fission products that are stable as chlorides under the conditions in the electrorefiner (e.g., cesium, strontium, barium).

The metallic products associated with all three electrodes also contain entrained electrolyte salt and cadmium.

- The cathodes are separately inserted into a cathode processor in which the entrained electrolyte salt and cadmium are recovered for recycle by vacuum distillation.

- The uranium metal is converted to an appropriate form, either hexafluoride for reenrichment or oxide for direct reuse or disposal. The extent to which additional cleanup of the uranium might be necessary before conversion is not known.

- The TRU metal goes to an injection casting furnace (not shown) where it is refabricated into new fuel for a fast transmutation reactor.

- The metal left at the anode, including the cladding, is heated in a metal waste furnace to produce a solid metallic waste form having zirconium as the major constituent for LWR fuels and iron as the major constituent for stainless-steel clad fuels.

- The fission-product-laden salt is circulated through a zeolite ion exchange bed which incorporates the salt and fission products into the zeolite matrix. The loaded zeolite is consolidated into a monolithic form by combining it with borosilicate glass frit and sintering it, which converts the zeolite to the mineral sodalite in a waste form called glass-bonded zeolite [NAS, 2000] [Kim, 2006]. Processes to remove fission products from the salt and recycle the salt are under development [Simpson, 2007].

An important obstacle to widespread adoption of pyroprocessing is that reprocessing is currently being carried out worldwide using aqueous processes and a very large experience base exists in large, well-established PUREX process plants. Consequently, there has been little demand for pyrometallurgical or other systems.

Although the technology for pyroprocessing SNF is not as advanced as that for aqueous reprocessing, a number of important differences between the two types of processes are evident, as summarized below:

- Pyroprocessing is inherently a batch process which means that materials must be moved as solid physical objects among most of the various steps described above. The size of the batches is limited by criticality considerations. The maximum throughput of a single electrorefiner is about 50 MTIHM/yr [GE-H, 2007]. On this basis, a pyroprocessing plant would require the operation of 16 electrorefiners in parallel to achieve the 800 MTIHM/yr throughput of the French UP3 aqueous reprocessing facility. The large number of movements of highly radioactive objects containing fissile materials in this manner is likely to require high equipment reliability, low accident likelihood, and a greater need for nuclear material accountability.

- Pyroprocessing will generate a somewhat different matrix of wastes and effluents:

  – The two highest-activity waste streams (sintered zeolite and metal waste ingot) are conceptually similar to wastes from aqueous reprocessing, but their characteristics are likely to be different.

  – Processes to generate a concentrated stream of cesium/strontium for separate management have not yet been developed.

  – Iodine becomes chemically combined with the molten pyroprocessing salts so there is no separate iodine recovery or waste form. The behavior of iodine during molten salt cleanup by zeolites must be determined.

  – Carbon is reduced to graphite in the process. It is unclear how it will be recovered or managed.

  – There is no estimate of the amount and characteristics of failed or used equipment such as electrodes and crucibles.

  – The use of cadmium indicates the potential for mixed wastes, but the extent to which this might occur is unknown.

  – Waste streams are solid under ambient conditions which avoids the need for large liquid waste storage tanks.

- Pyroprocessing per se does not use organic chemicals. This avoids the potential for accident scenarios involving organic chemical reactions (e.g., fire, red oil, resin explosions) and wastes from cleanup of organic solvents and extractants.

- The chemicals used in pyroprocessing can tolerate extremely high levels of radiation without unacceptable degradation which allows high burnup, short-cooled SNF to be reprocessed.

- Pyroprocessing yields a TRU product containing significant concentrations of fission products in a single process step in a closed vessel, which is advantageous from the standpoint of proliferation prevention. However, pure plutonium could still be recovered from the TRU product using other processes, which may include altering the operation of the pyroprocessing system.

- After repeated batch processes, the salt accumulates impurities and must be discarded.

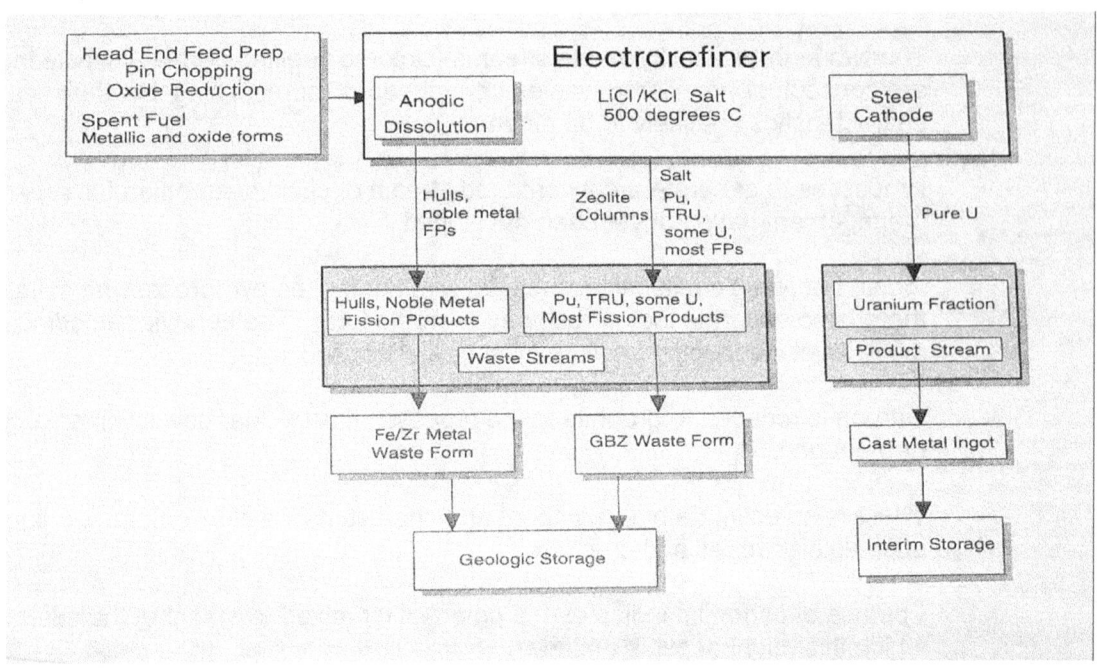

Figure 20: Schematic diagram of pyroprocessing with uranium recovery

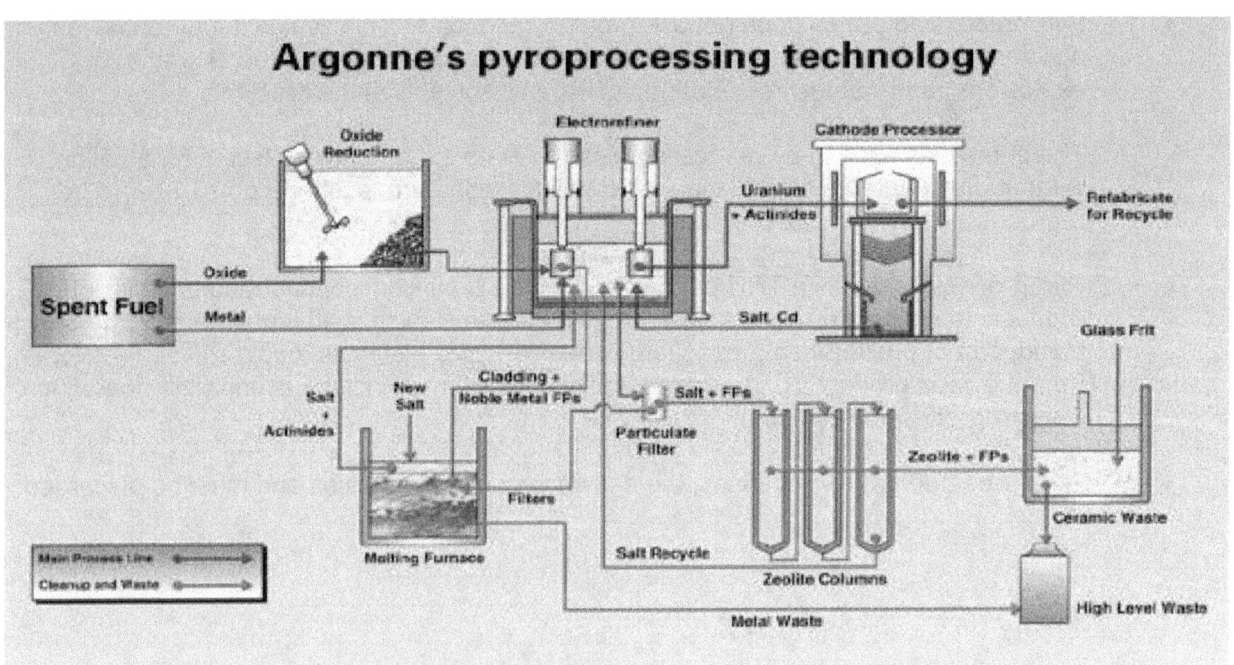

Figure 21: Pictorial representation of pyroprocessing operations

6.3. Reprocessing High-Temperature Gas-Cooled Reactor Fuels

HTGR fuels are distinctly different from other reactor fuels. This difference imposes a very different type of head-end processing. Unlike most other reactors, the HTGR fuel is not a ceramic oxide fuel clad in metal tubing. HTGR fuel is made mostly of graphite and is in one of two geometric configurations, the spherical (pebble) form, and the prismatic form mentioned above, both of which are unlike any other reactor fuels. (See the discussion of the composition of HTGR fuels in Section 2.2.4) There has been no commercial reprocessing of HTGR fuels, although development work has been conducted at ORNL and elsewhere. Some of the salient features of HTGR fuel reprocessing are discussed below.

6.3.1. Flowsheets

The first step in reprocessing HTGR fuels is removal of the bulk of the graphite, whether it is in the form of balls or prismatic blocks [Del Cul, 2002]. Several removal approaches have been considered. The balls would be crushed or burned to release the TRISO particles, which contain the fuel material of interest. The crushed material would be sieved to recover the fuel particles, and the inert graphite would become a waste stream. The separated fuel particle would then be put through a grinder to break the TRISO coatings and expose the tiny fuel kernels that contain the uranium and actinides and fission products. Finally, the crushed fuel material and any residual graphite would be dissolved in nitric acid preparatory to solvent extraction. Alternatively, the residual graphite could be burned either before or after crushing the fuel. The advantage of early removal of the graphite by crushing or burning is that it would remove the bulk of the graphite before dissolution in nitric acid. Nitric acid dissolution of finely ground graphite and carbides produces organic compounds that could interfere with the solvent extraction separation step, which is the next step in reprocessing. In any case, the fragments of the SiC inner coating would need to be removed before the solvent extraction step, because their presence could interfere with the operation of the solvent extraction equipment, especially if centrifugal contactors were used.

For the prismatic fuel blocks, it is desirable to separate the coated microspheres from the bulk of the graphite block as a first head-end step. This might be done by burning, as described above, or reaming the carbonized fuel sticks out of the blocks. In this way, the bulk of the graphite could be physically removed, leaving the coated microspheres for treatment as outlined above for the fuel balls. The de-fueled prismatic blocks could then be disposed of in the same way that graphite from reactors is managed [IAEA, 2006; Wickham, 1999] (i.e., by permanent removal from the environment as solid graphite, destruction [e.g., incineration and recycling]).

6.3.2. Unusual Plant Features

The head end of the HTGR spent fuel reprocessing plant would have unique features arising from the necessity to crush, grind, or burn the graphite fuels. These steps are to be contrasted with the relatively much simpler fuel shearing employed with LWR fuels. After these head-end steps, the remainder of the plant would be essentially conventional solvent extraction using PUREX or other suitable process, assuming that interference from organic compounds formed by reaction of nitric acid with graphite could be kept acceptably low.

6.3.3. Reprocessing Wastes

The bulk of the graphite would become a moderately radioactive waste. The radioactivity would primarily result from failed fuel particles that could release small amounts of radionuclides into the pebbles or the prismatic blocks, but it would also contain amounts of ^{14}C that are large compared to what is in the fuel matrix. In the case where the graphite is burned, there would be a CO_2 gaseous waste. Volatile radionuclides would be trapped in the off-gas filters or subsequent trapping systems. The number and types of wastes from the separation processes would depend on the processes chosen, and on whether the fuel was based on the uranium-plutonium or the uranium-thorium fuel cycle. However, if the present UREX+1a flowsheet were used, the wastes should be similar to those from processing LWR fuels with the exception of (1) producing much more ^{14}C in the form of CO_2 or a solid ^{14}C waste form and (2) generating a waste stream of SiC hulls in lieu of metal hardware.

6.4. French Proposals

6.4.1. GANEX

The French have been especially active in pursuing a variety of proliferation-resistant reprocessing methods [Boullis, 2006] other than PUREX. The CEA has developed the GANEX (grouped actinide extraction) process. It is designed to reduce the radiotoxicity and heat output of final wastes. It is envisaged for possible adoption at the La Hague plant in about 2040 [Cazalet, 2006]. The GANEX process makes no attempt to separate anything but the actinides and lanthanides as a group from most of the uranium and then from each other. Cesium and strontium remain with the fission products.

In the GANEX process, shown in very simplified form in Figure 22 [Bouchard, 2005], uranium is separated in a preliminary step and the raffinate then undergoes three subsequent extractions, which result in an actinide stream which is combined with the uranium product from the first step. The lanthanides and other fission products, including cesium, strontium, and technetium, are formed into borosilicate glass for storage and deep geologic disposal.

The GANEX process has the disadvantage of leaving the high heat-emitters cesium and strontium and potentially mobile technetium with the other fission products in the vitrified waste glass destined for disposal. It is a modest extension of the PUREX process which could likely be implemented with little or no additional R&D concerning the central processes. However, significant additional development of waste processing and treatment technologies would likely be needed to meet U.S. requirements.

GANEX
Process

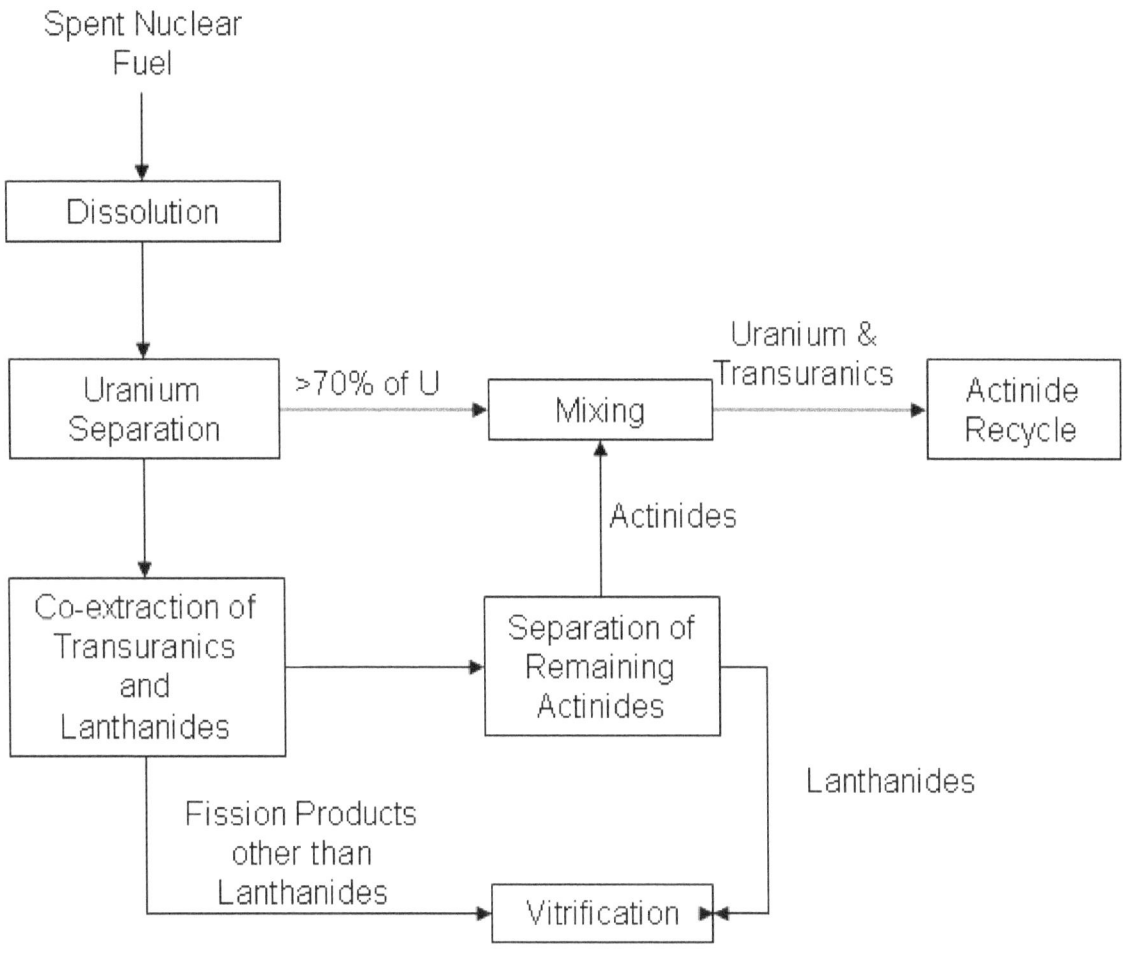

Figure 22: The French GANEX process

6.4.2. COEX™

The COEX™ (co-extraction) process, which was developed by CEA and AREVA, is divided into the three main phases shown in Figure 23 and described as follows:

- Extraction cycles for separating and purifying a uranium-plutonium mixture without ever isolating pure plutonium have the following steps:

    - SNF is dissolved in nitric acid. The dissolver solution is contacted with TBP extractant in an organic diluent to recover the uranium and plutonium while the fission products and minor actinides remain in the nitric acid solution. The fission products and minor actinides are concentrated by evaporation and then vitrified. The extractant still contains some residual fission products and minor actinides in addition to the uranium and plutonium. The minor actinides are separated from the uranium-plutonium mixture a nitric acid washing process.

    - The uranium and plutonium are separated into two streams: a uranium stream and a mixed uranium-plutonium stream.

    - The uranium-plutonium mixture is purified by another solvent extraction cycle extraction.

- The uranium-plutonium nitrate solution is converted to $(U,Pu)O_2$ by first adding a quantity of uranium nitrate to adjust the solution to the required concentration. The uranium-plutonium nitrate solution is brought into contact with oxalic acid which simultaneously precipitates the U-Pu as the oxalate. The precipitate obtained is then filtered, dried and calcined to form a homogeneous uranium-plutonium oxide powder.

- Fresh MOX fuel is manufactured using a powder metallurgy process similar to that described in Section7.

The COEX™ process is a modification of the PUREX process which could likely be implemented with little or no additional R&D concerning the central processes. However, significant additional development of waste processing and treatment technologies may be needed for COEX™ or other reprocessing flowsheets if U.S. requirements differ significantly from those in other countries such as France and the UK.

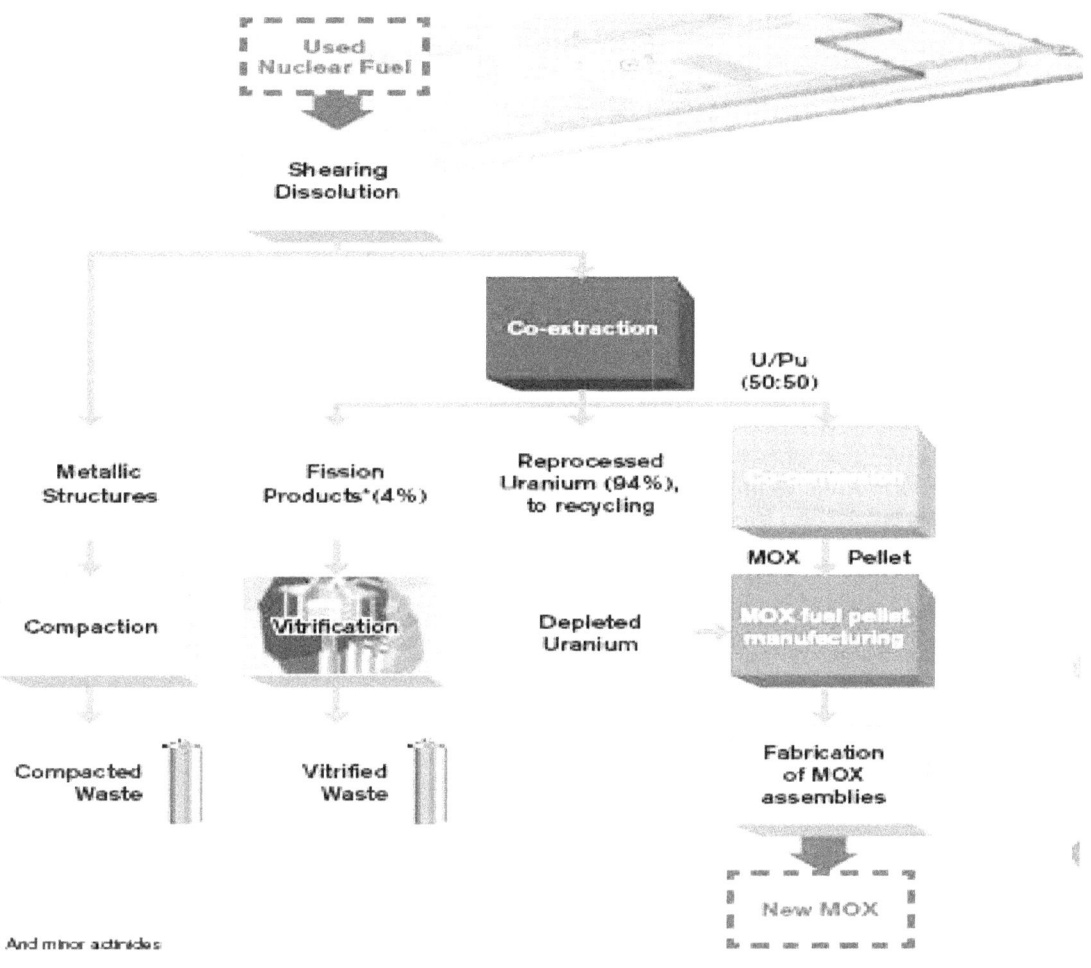

Figure 23: The French COEX™ process

6.5. General Electric's Pyroprocess

GE has proposed [Loewen, 2007] a path to deploy the GNEP CFTC based on the ANL pyrometallurgical process that GE reported to have extensive testing not only in the United States, but also in Russia, Japan, and South Korea. The proposed process is based on a modular concept that would be sized to support a fast transmutation reactor for actinide burning. It is claimed to be proliferation-resistant and to have a low environmental impact. The process would be operated either batchwise or continuously.

Although the pyrometallurgical process is best suited to spent metallic fuels, as noted above, it could be adapted to oxide fuels through the use of cathodic or carbon reduction of the oxide in a molten LiCl at 650 °C to produce metal. The oxygen or CO_2 would be released. This reduction has been demonstrated at ANL at a kilogram scale. GE plans to demonstrate the electro-reduction operation at its Wilmington, South Carolina, plant using the current SNM license and then to license a site using lessons learned at the Wilmington plant.

7. ADVANCED FUEL REFABRICATION

Current preparation of conventional pelletized reactor fuels for LWRs and fast reactors (see Section 3.2) requires grinding to achieve specified size and shape. This process produces finely divided fuel particles that must be recovered and recycled. A "dust-free" sol-gel microsphere pelletization process has been developed for fabrication of $(U,Pu)O_2$, $(U,Pu)C$, and $(U,Pu)N$ fuel pellets containing around 15 percent plutonium [Ganguly, 1997]. The microspheres can be pressed into pellets that can be sintered to 85 percent of theoretical density in $Ar+8\%$ H_2 at 1700 °C. The sintered oxide, monocarbide, and mononitride pellets have an open pore microstructure with fine grain size.

Hydrated gel-microspheres of UO_3/PuO_2 and $UO_3/PuO_2/C$ are prepared from nitrate solutions of uranium and plutonium by the "ammonia internal gelation" process, using hexamethylene tetramine that decomposes to ammonia in the presence of silicone oil at 90 ± 1 °C in a gelation bath. For oxide fuel pellets, the hydrated UO_3/PuO_2 gel-microspheres are calcined at around 700 °C in $Ar/8\%$ H_2 atmosphere to produce "non-porous," "free-flowing," and coarse (around 400-micron) microspheres which are directly pelletized at 550 megapascals to green pellets. The MOX pellets are subjected either to low temperature (about 1100 °C) oxidative sintering in N_2/air containing about 1500 ppm O_2 or to high temperature (about 1650 °C) sintering in $Ar/8\%$ H_2.

For monocarbide and mononitride pellets, hydrated gel-microspheres of $UO_3/PuO_2/C$ were subjected to carbothermic synthesis in vacuum (about 1 pascal and flowing nitrogen (flow rate of 1.2 m^3/h) in the temperature range of 1450–1550 °C. The microspheres retain their individual identity in the sintered pellets because, during sintering, densification takes place mainly within and not between the microspheres.

Metallic fuels of uranium/plutonium/zirconium continue to be of considerable international interest because of their very promising performance at high fuel burnup in fast reactors.

8. REGULATION AND LICENSING OF FUEL RECYCLE FACILITIES

8.1. Licensing—A Historical Perspective

Application of the NRC's regulatory process to commercial recycle facilities will not be simple. Deficiencies in regulations are known to exist that will require new rulemaking or many exemptions before a license can be approved (for example, for reprocessing SNF at a commercial site). The body of regulations that currently establishes the NRC's licensing and regulatory process for recycle facilities, associated waste streams, and effluents include at least the following requirements in Title 10 of the *Code of Federal Regulations*:

- Radiation Protection— Part 20
- Uranium Product Conversion—Part 40
- Reprocessing— Part 50
- Environmental Protection— Part 51
- Operator's Licenses – Part 55
- Low-Level Waste Disposal—Part 61
- Disposal of HLW at Yucca Mountain—Part 63
- Fuel Fabrication—Part 70
- HLW Vitrification and Storage—Part 70
- Plutonium Product Conversion—Part 70
- Reprocessed Uranium Storage—Part 70
- Transportation—Part 71
- Spent Nuclear Fuel Storage—Part 72
- Physical Protection—Part 73
- Material Control and Accountability—Part 74
- Cesium/Strontium and TRU Storage—Part 30 and Part 70
- Decommissioning—Part 50 and Part 51
- Licensing Process— Part 50/52 and Part 70

This chapter focuses on SNF reprocessing because there has been significant experience with licensing fuel fabrication plants. However, only limited regulatory experience exists for licensing and regulating reprocessing facilities. Most of this experience occurred decades ago under the AEC and the then newly formed NRC.

8.1.1. Licensing Experience at Nuclear Fuel Services

In 1966, the AEC used 10 CFR Part 50 to license the NFS reprocessing facility at West Valley. From 1966 to 1972, NFS reprocessed 640 MTIHM of fuel at West Valley, but in 1972 the facility shut down to implement a number of improvements and never restarted. Since that time, the NRC has not approved any other licenses for reprocessing SNF, although BNFP had been undergoing a licensing review when President Carter terminated commercial reprocessing. Although some 30 years have passed, 10 CFR Part 50 still remains the default licensing basis for reprocessing SNF. Many changes have occurred to 10 CFR Part 50 during that period, but most relate to licensing utilization or power reactor facilities as opposed to reprocessing facilities like the one at West Valley.

8.1.2. Licensing Experience at Barnwell

The Preliminary Safety Analysis Report for the BNFP Separations Facility was submitted on November 6, 1968. Following appearances before ACRS and a public hearing before an Atomic Safety and Licensing Board, the AEC issued a construction permit on December 18, 1970. Subsequently, the applicant submitted several substantial documents to the AEC, including the Environmental Report and Facility Safety Evaluation for the Uranium Hexafluoride Facility, Updated Environmental Report for the Separations Facility, Final Safety Analysis Report for the Separations Facility (five volumes and several addenda), Technical Description in Support of Application for FRSS Operation, Preliminary Safety Analysis Report for Plutonium Product Facility, and Nuclear Materials Safeguards Supplement. In addition, the applicant submitted many documents containing responses to AEC questions.

To comply with the National Environmental Policy Act, a public hearing took place before an Atomic Safety and Licensing Board preliminary to the issuance of an operating license. The AEC verified the compliance with applicable regulations and the commitment implicit in issuing the construction permit and conducted more than 20 formal inspections before the cessation of commercial licensing activities (brought about by Presidents Carter and Ford).

Some facilities and operations of the plant were being licensed under 10 CFR Part 50 regulations, while other facilities, such as the plutonium nitrate conversion plant, were being licensed under 10 CFR Part 70. The BNFP licensing process was complicated by the evolving character of regulations pertaining to reprocessing plants and waste management, and the interrelation between the licensing of the facility and other regulatory actions taking place concurrently. Notable among the latter were the proceedings on the Generic Environmental Statement on Mixed Oxide Fuel (GESMO) and the Environmental Statement on the Management of Commercial High-Level and Transuranic Contaminated Radioactive Waste. These latter activities, however, were placed on hold when the International Nuclear Fuel Cycle Evaluation (INFCE) was invoked.

8.2. Current Licensing Process and Alternatives

Under current regulations, both production (reprocessing) and utilization facilities (power reactors) must comply with 10 CFR Part 50 to obtain a construction or operating license. To ensure adequate protection of public health and safety, 10 CFR Part 50.34 requires that applicants must demonstrate that their designs meet general design criteria and mitigate a postulated set of accidents known as "design-basis" accidents to within certain specified radiological release limits. Applicants use plant-specific PRA insights to ensure that the plant is protected against a robust set of accidents (although this is not required under 10 CFR Part 50).

Because 10 CFR Part 50 was not written specifically for reprocessing SNF, there are deficiencies in its use. For example, 10 CFR 50.20, "Two Classes of Licenses," does not specifically acknowledge the licensing of reprocessing plants, and paragraph (a) of 10 CFR 50.34, "Contents of Applications; Technical Information," is directed solely to reactors. The National Environmental Policy Act processes, that require documentation for a reprocessing facility, have yet to be completely demonstrated. Earlier efforts in this regard for commercial reprocessing plants occurred after the submission of the safety analysis report and thus were very time consuming and contentious. Modification of the current 10 CFR Part 50, or exemption to its requirements, would be needed to accommodate the technical differences between licensing LWRs and reprocessing facilities. Rule modifications could be extensive, and public hearings on exemptions are likely to result in a lengthy process.

All fuel fabrication facilities are licensed under 10 CFR Part 70, "Domestic Licensing of Special Nuclear Material," Subpart H, "Additional Requirements for Certain Licensees Authorized To Possess a Critical Mass of Special Nuclear Materials." Experience and lessons learned from licensing fuel fabrication facilities under 10 CFR Part 70 are to some extent applicable to reprocessing facilities. The regulation utilizes an ISA, sometimes known as a process hazards analysis, to assess the safety of the design and to identify equipment relied on for safety. Use of ISA is an important step towards risk quantification and expanded use of risk-informed regulations. However, in a January 14, 2002, letter to the Commission [ACNW&M, 2002], the Joint Subcommittee of ACRS and ACNW noted shortcomings in ISAs that would likely need to be addressed to expand its role in regulatory decisions involving reprocessing facilities. Additionally, measurable limits on emissions from refabrication facilities would need to be established and implemented to ensure public and environmental protection.

At the time of this writing, a new rule (10 CFR Part 53, "Risk-Informed, Performance-Based Framework") is under development. This rule is expected to provide a risk-informed, performance-based framework for licensing the next generation of nuclear reactor designs. The framework integrates safety, security, and emergency preparedness to establish a comprehensive set of requirements as a license condition. The approach focuses on the most risk-significant aspects of plant operations and uses the Commission's safety goals (separate goals would need to be developed for recycle facilities) as top-level regulatory criteria that designers must meet to ensure adequate safety. The approach eliminates the need for exemptions by implementing guidance to accommodate technological differences between designs. Such an approach to licensing reprocessing facilities may be advantageous because of its flexibility. However, 10 CFR Part 53 is primarily intended for new commercial power reactors, and its current schedule may not support its application to reprocessing facilities.

In addition to the modification of existing regulations, a new rule could be designed specifically for licensing recycle facilities. A new rule could avoid the need to write exemptions for rules already in place and would place all the regulations relevant to the recycle facilities under one part of the regulations, effectively leaving other parts of the regulations unchanged. The rule could be made to expedite the licensing process by eliminating exemptions and protracted hearings. The drawback is that developing such a rule is likely to require extensive resources and time, although it is unclear whether the requirements are significantly greater than those of other approaches.

It is expected that implementation of any new or modified rule would be consistent with Commission policies including the Commission's PRA policy statement [NRC, 1995], which states in part, "The use of PRA technology should be increased in all regulatory matters to the extent supported by the state of the art in PRA methods and data, and in a matter that complements the NRC's deterministic approach and supports the NRC's traditional defense-in-depth philosophy." The Committee has gone on record repeatedly in letters to the Commission about the use of risk-informed decision making, starting in October 1997 and most recently in the letter of May 2, 2006. These letters are listed in Appendix C. Additionally, ALARA requirements for reprocessing facilities that establish design objectives and limiting conditions for radioactive material effluents (analogous to the current Appendix I, "Numerical Guides for Design Objectives and Limiting Conditions for Operation to Meet the Criterion 'As Low as Is Reasonably

Achievable' for Radioactive Material in Light-Water-Cooled Nuclear Power Reactor Effluents," to 10 CFR Part 50) will need to be formulated.

8.3. Environmental Protection

Production and utilization facilities must comply with environmental protection regulations. Both (1) designed-in barriers that block the release of radioactive material to the environment, and (2) operational performance and characteristics that limit the release of radioactive material to the environment provide protection.

8.3.1. Design Perspective

Before facility construction, 10 CFR Part 51, "Environmental Protection Regulations for Domestic Licensing and Related Regulatory Functions," requires that each applicant submit an environmental report that complies with Table S-3, "Table of Uranium Fuel Cycle Environmental Data," at 10 CFR 51.51(b), as the basis for evaluating the contribution of its activity to the environment. Currently, Table S-3 considers only two fuel cycles, uranium-only recycle and no recycle. To accommodate recycle of plutonium and other actinides, the NRC staff would need to consider whether Tables S-3 and S-4, "Environmental Impact of Transportation of Fuel and Waste to and from One Light-Water-Cooled Nuclear Power Reactor," at 10 CFR 51.52(c) encompass the releases of radioactive waste to the environment. This consideration would address the impact of fuel recycle on environmental aspects listed in Table S-3 other than the release of radioactivity (e.g., land [temporarily committed], occupational exposure, water discharges). The staff would need to reconsider Table S-4 to determine if it encompasses the environmental impact of transportation of fuel and radioactive waste, with consideration of the changes introduced by the recycling of SNF.

An applicant for a fuel recycling facility would need to provide an environmental report with the information required by 10 CFR 51.45, "Environmental Report." The applicant would need to develop and provide information on the other stages of the fuel cycle analogous to the information in Tables S-3 and S-4, which apply to LWRs. Establishing a new set of tables analogous to S-3 or S-4 explicitly for reprocessing facilities may be the best approach should reprocessing become a mainstream activity. 10 CFR 51.45(d) requires an environmental report that discusses the status of compliance with the applicable environmental quality standards and other requirements including those imposed by Federal, State, regional, and local government agencies.

The U.S. Environmental Protection Agency (EPA) standard (40 CFR Part 190, "Environmental Radiation Protection Standards for Nuclear Power Operations") places limits on the entire fuel cycle, and applicants seeking to design a facility would need to comply. Additionally, this standard does not encompass plutonium or actinide refabrication and reuse in a reactor, and it would appear necessary do so if SNF recycle were to proceed.

Following the receipt of the environmental report, the applicant will need to prepare an environmental impact statement (EIS) for siting, construction, and operation of the recycle plant(s). This requirement can be very time consuming and contentious and must be started well in advance of planned plant construction. The design of a spent fuel reprocessing plant, for example, is dictated to a large extent by the requirements to (a) protect the plant operators from radiation, to provide a safe working environment, and to prevent criticality and (b) limit routine and accidental releases of radionuclides to the public.

The design must be such that the protection provided by the radiation shielding and confinement of radioactivity keeps radiation doses ALARA and consistent with the allowable limits of personnel dose (10 CFR Part 20, "Standards for Protection against Radiation"), and air and water contamination (Appendix I to 10 CFR Part 50). Exceptions to the dose limits may be made in the case of planned special exposures, but in any case, the ALARA principle applies.

8.3.2. Operating Perspective

In 1974, the AEC initiated a study to examine the environmental impact from SNF reprocessing and widespread use of MOX fuel as a means to use the uranium and plutonium products of reprocessing. The study, often referred to as GESMO, was published in August 1976 [NRC, 1976] and considered five alternatives:

(1)   prompt fuel reprocessing, prompt uranium recycle, delayed plutonium recycle
(2)   delayed fuel reprocessing, followed by uranium and plutonium recycle
(3)   prompt uranium and plutonium recycle
(4)   uranium recycle; no plutonium recycle
(5)   no uranium or plutonium recycle

Findings from the study found no clear preference for any of the alternatives. Differences in health effects between the fuel cycles did not provide a basis for choosing one approach over another. Any environmental benefit that did result arose from the conservation of uranium resources and not from differences in the management of radiological waste.

The study found, however, that for the various recycle alternatives (as opposed to the once-through fuel cycle), the collective dose increased by several orders of magnitude. Three radionuclides that were assumed not to be removed from the reprocessing plant gaseous effluent contributed to this dose (^{3}H, ^{14}C, and ^{85}Kr). The contribution of ^{129}I and ^{131}I was much smaller because most of the iodine was assumed to be removed from the gaseous effluent.

Although the dose to any one individual was found to be small, the large integrated (world) population exposed to the gaseous effluents drove the results. While public hearings were being held on the GESMO study and BNFP license, the Carter Administration terminated reprocessing in the United States. The public hearings were never completed, and the Commission postponed its decision on whether to allow the wide-scale use of MOX fuel in LWRs. This could become an issue once again should wide-scale reprocessing be considered as a mainstream activity.

In about the same time period as the GESMO study (January 13, 1977), EPA released 40 CFR 190, Subpart B, which established the environmental standard for the complete uranium fuel cycle. The rule prescribed two criteria:

- Subpart B, 40 CFR 190.10(a): Annual dose equivalent to any member of the public for the entire fuel cycle

    – Whole body   <25 mrem
    – Thyroid   <75 mrem
    – Any other organ   <25 mrem

- Subpart B, 40 CFR 190.10(b): Radioactive material released to the environment per GWe energy produced

  - ^{85}Kr <50,000 curies
  - ^{129}I <5 millicuries
  - Pu and other alpha-emitting radionuclides with half-lives >1 year   <0.5 millicuries

EPA set December 1, 1979 as the effective date of 40 CFR 190.10(a), except for operations associated with uranium milling which were given an effective date of December 1, 1980. The Agency also established the effective date for 40 CFR 190.10(b) as December 1, 1979, except for ^{85}Kr and ^{129}I standards which became effective on January 1, 1983. Soon after the standards were released, stakeholders expressed concerns that the standards were overly conservative, costly, and would require technology that was considered beyond the state of the art. In addition, industry believed that requirements should not be established until international agreements were reached that would restrict emissions from foreign sources. Nevertheless, EPA approved part of the standard (except for ^{85}Kr) on December 1, 1979, and an ^{85}Kr standard that became effective on January 1, 1983. By that time, all reprocessing activities had ceased, and interest in the new standard vanished.

Today, the EPA standard for utilization (power reactor) facilities is being met through the NRC's enforcement of Appendix I to 10 CFR Part 50, which sets the following operating limits:

- Liquid Effluents   <3 mrem whole body, or <10 mrem to any organ

- Gaseous Effluents   <5 mrem whole body, or <15 mrem to the skin

- Radioactive Iodine   <15 mrem to any organ and other material in particulate form in effluents to the atmosphere

These limits provide the basis for implementation of EPA standards for single reactor units. The remaining (nonutilization) portion of 40 CFR Part 190 release is divided among the rest of the fuel cycle which to date has not included reprocessing.

8.4.   Decommissioning

Decommissioning commercial reprocessing plants can be very costly. Information based on decommissioning experience is limited because so few reprocessing plants have been decommissioned.

In 1976, NFS withdrew from the reprocessing business and turned control over to the site owner, the New York State Energy Research and Development Authority. In 1980, the NRC suspended West Valley's license to reprocess SNF, and the West Valley Demonstration Project Act was executed to clean up the site and its facilities. Under the Act, the NRC retained certain responsibilities including prescribing decontamination and decommissioning criteria.

Significant lessons learned and regulatory actions have resulted from the West Valley decommissioning experience. The cost to clean up the site to date has exceeded $2 billion, although a fund of only $4 million had been set aside for decommissioning and decontamination.

The English Sellafield reprocessing plant is currently undertaking decommissioning of its "first generation" reprocessing plants, including analytical laboratories, legacy wastes, and the "North Compound," a facility established to support Windscale pile operation and subsequently extended to include waste storage. The total estimated cost of this interim decommissioning is about £20 million (equivalent to about $40 million) [Sellafield, 2005].

The French UP1 reprocessing plant at Marcoule has an estimated decommissioning cost of €5.6 billion (about $7.6 billion), about half of which is for treating wastes stored on site [Hore-Lacey, 2007]. Thus, there is a very wide range of real and anticipated reprocessing plant decommissioning costs. Additional decommissioning experience and the incorporation of decommissioning lessons learned into future plants will result in better estimates of the costs for future reprocessing plant decommissioning.

Shortfalls in decommissioning funding like that at West Valley resulted in Appendix F to 10 CFR Part 50. Sections of Appendix F that are relevant to reprocessing plants include the following:

- Paragraph 3, which states, "Disposal of high-level radioactive fission product waste material will not be permitted on any land other than that owned and controlled by the Federal Government."

- Paragraph 2, which states, "High level liquid radioactive wastes shall be converted to a dry solid as required to comply with this inventory limitation, and placed in a sealed container prior to transfer to a Federal repository in a shipping cask meeting the requirements of 10 CFR 71. Upon receipt, the Federal Repository will assume permanent custody of the waste materials although industry will pay the Federal Government a charge which together with interest on unexpended balances will be designed to defray all costs of disposal and perpetual surveillance."

- Paragraph 5, which states, "Applicants proposing to operate fuel reprocessing plants, in submitting information concerning financial qualifications as required by Section 50.33(f), shall include information enabling the Commission to determine whether the applicant is financially qualified, among other things, to provide for the removal and disposal of radioactive wastes, during operation and upon decommissioning of the facility."

Although Appendix F may reduce the likelihood of shortfalls in decommissioning funding, the Commission, in a staff requirements memorandum (SRM) dated February 7, 2006 [NRC, 2006c], directed that an important design criterion for any new reprocessing effort will be that decommissioning costs be manageable. NRC guidance under development should help designers address these concerns at the conceptual design stage.

Any new license application must also address how the design and procedures for operating the facility will minimize contamination of the facility and the environment and facilitate eventual decommissioning (10 CFR 20.1406). This includes realistic estimates of the funds required for decommissioning (10 CFR 72.30, "Financial Assurance and Recordkeeping for Decommissioning"), including site characterization, cleanup, waste disposal, and surveillance.

The NRC Office of Nuclear Regulatory Research is developing a regulatory guide to implement 10 CFR 20.1406 to facilitate decommissioning.

Additional NRC regulatory requirements related to decommissioning include the following:

- 10 CFR Part 20, Subpart E, "Radiological Criteria for License Termination"

    – 10 CFR 20.1402, "Radiological Criteria for Unrestricted Use"

    – 10 CFR 20.1403, "Criteria for License Termination under Restricted Conditions"

    – 10 CFR 20.1404, "Alternate Criteria for License Termination"

    – 10 CFR 20.1405, "Public Notification and Public Participation"

    – 10 CFR 20.1406, "Minimization of Contamination"

- 10 CFR Part 72, "Licensing Requirements for the Independent Storage of Spent Nuclear Fuel and High-Level Radioactive Waste and Reactor Related Greater Than Class C Waste"

    – Subpart B, 10 CFR 72.30

    – Subpart D, 10 CFR 72.54, "Expiration and Termination of Licenses and Decommissioning of Sites and Separate Buildings or Outdoor Areas"

9. ISSUES ASSOCIATED WITH LICENSING AND REGULATING FUEL RECYCLE FACILITIES

The focus of this chapter is on licensing and regulation of industrial-scale fuel reprocessing and refabrication facilities. As suggested by the foregoing information, a number of licensing or regulatory issues warrant consideration before receipt of a license application. The following sections identify these issues and offer insight into their resolution.

9.1. Selection or Development of Licensing Regulation(s) for Recycle Facilities

A key issue to be decided before receipt of a license application for SNF recycle facilities is what primary regulation(s) should be used to license each facility and what approaches (e.g., probabilistic versus deterministic safety assessments) should be used to develop a new regulation or modify an existing regulation. For the purposes of this paper, the authors assumed that the specific regulations and approaches used to license well-established fuel cycle facilities and operations (e.g., interim storage of spent fuel, radioactive material transportation, uranium fuel fabrication, reactors) will not change.

The novel facilities that will necessitate decisions concerning the appropriate licensing regulations and approaches include the following:

- reprocessing fuels from LWRs and later from other advanced reactors

- fabrication of fuels to recycle TRU or fission product elements or fuels for some new reactor designs (e.g., graphite-moderated reactors)

- disposal of new types of wastes such as cladding and TRU (GTCC) waste

- extended interim storage of intermediate-lived radionuclides (cesium and strontium), followed by in situ disposal.

The following sections discuss some of the factors that should be considered when making these decisions.

9.1.1. Multiple Regulatory Paths Available

As discussed in Section 8, there are a number of existing regulations, as well as the possibility of developing one or more entirely new regulations, for licensing recycle facilities. None of the existing regulations is entirely suitable for the fuel recycle facilities. While Section 8.2 gives detailed reasons for this, the overarching reason is that existing regulations were designed for (1) reactors where maintaining heat removal capability in situations involving fast transients in a core with a high-power density is an important purpose of the regulations but where there are modest chemical hazards and few concentrated solutions of radionuclides, or (2) facilities that handle relatively small amounts of radioactivity because they process only uranium. Maintaining the fast-response capability to remove large amounts of decay heat is not particularly important in fuel recycle facilities, but there are substantial amounts of radioactivity in fluids and a higher likelihood of inadvertent criticality, in addition to a variety of toxic and potentially flammable or reactive chemicals in routine use. These differences lead to the need for substantial modification of existing regulations or development of new regulations directed at particular types of facilities to address the specifics of fuel recycle facilities.

9.1.2. NRC Staff's Proposed Options and Commission Direction for Licensing GNEP Recycle Facilities

The NRC staff [NRC, 2007a] identified the four options summarized in Table 20 for developing a regulatory framework to license advanced reprocessing and burner reactor facilities.

Table 20: <u>Regulatory Options for Advanced Fuel Recycle and Burner Reactor Facilities</u>

Option	CFTC	ABR
1	Revise 10 CFR Part 70 to include spent fuel reprocessing; consider additional safety analysis requirements for a reprocessing facility; and revise 10 CFR Part 50 as appropriate.	Use existing 10 CFR Part 50, with exemptions as necessary, or a suitably modified and adapted 10 CFR Part 52 process, to address sodium-cooled fast reactor technology.
2	Same as Option 1.	Create a new regulation specific to advanced recycling reactors (10 CFR Part 5X).
3	Develop a specific GNEP regulation applicable to fuel reprocessing, refabrication, and recycle reactors (10 CFR Part XX).	
4	• Issue a *Federal Register* notice in FY 2007 soliciting public and stakeholder input on desirable attributes of the regulatory framework for GNEP, as well as comments on whether there are any major substantive technical issues relating to an accelerated schedule that may affect development of GNEP regulations and/or how such facilities should be regulated. • After consideration of public and stakeholder comments, decide either to issue an order or direct a rulemaking to establish specific requirements. • Concurrently, develop a licensing-basis document for fuel separations/fuel fabrication/advanced recycling reactor facilities for use by the Commission in developing an order or as the technical basis for the rulemaking process, as appropriate.	

The NRC staff's options are similar to those described in Section 8.2 of this paper. After evaluating the pros and cons for each of the options, the staff recommended that the Commission proceed with Option 1 in a phased approach. The first phase would involve development of the regulatory framework by preparing technical basis documents to support rulemaking for 10 CFR Part 70 (for fuel recycle facilities) and potential rulemaking for sodium-cooled fast reactors. The first phase would also involve exploration of whether 10 CFR Part 52, "Licenses, Certifications, and Approvals for Nuclear Power Plants," could be modified to address sodium-cooled fast reactors and a gap analysis of 10 CFR Part 50 to identify the changes in regulatory requirements that would be necessary to license recycle facilities and an ABR.

In the second phase, the NRC staff would shift to Option 3 and develop a new regulation for GNEP fuel recycle and reactor facilities. The analyses performed in the first phase would be used to evaluate whether unique programmatic or technical interrelationships exist among all closed fuel cycle technologies and could serve as a basis for developing a new regulation for advanced fuel recycle and burner reactor facilities.

In an SRM responding to the NRC staff's recommendations, the Commission [NRC, 2007b] directed the staff to begin developing the regulatory framework to license SNF recycle facilities using an option based on 10 CFR Part 70 by preparing the following:

- technical basis documentation to support rulemaking for 10 CFR Part 70 with revisions to 10 CFR Part 50 as appropriate to eliminate its applicability to licensing an SNF reprocessing plant

- a gap analysis for all NRC regulations (10 CFR Chapter I) to identify changes in regulatory requirements that would be necessary to license a reprocessing facility

The NRC has used 10 CFR Part 70 to license fuel fabrication facilities, and the regulation is currently the basis for reviewing the license application for the MOX fuel fabrication plant at the SRS. Experience and lessons learned from previous and ongoing use of 10 CFR Part 70 to license fuel fabrication facilities are likely to be useful when deciding how it should be modified to license SNF recycle facilities.

9.1.3. Important Factors in Developing Regulations for SNF Recycle Facilities

The NRC will need to consider the following important aspects of 10 CFR Part 70 and potential modifications to make the regulation efficient and effective for licensing SNF recycle facilities:

- Use of an ISA: 10 CFR Part 70 calls for the use of an ISA to evaluate the in-plant hazards and their interrelationship in a facility processing nuclear materials. Doses to the public are typically estimated using a scenario-based approach. Use of ISA is an important step towards quantifying risk as compared to traditional conservative, scenario-based deterministic approach. The primary reason for using ISA rather than full scope PRA is that the consequences of likely accidents in or routine releases from fuel cycle facilities are believed to be small compared to the consequences of accidents at reactors, and does not justify the effort of doing probabilistic analyses. However, the effort required to prepare an ISA for complex SNF recycle handling liquids containing substantial quantities of concentrated cesium, strontium, and TRU elements is likely to approach the effort that would be required to evaluate risks using a PRA. The Committee and the ACRS have previously advised [ACNW&M, 2002, 2006] that a regulation that utilizes PRA insights is preferable to one based on ISA because the latter has significant limitations in its treatment of dependent failures, human reliability, treatment of uncertainties, and aggregation of event sequences.

- Best estimate versus conservative: A companion issue to that of probabilistic versus deterministic approaches is whether analyses should be based on data and models that represent the best estimate of what might really occur with an associated uncertainty analysis to explore the effects of incorrect data or models, or should be based on demonstrably conservative data and models. Most regulations and license applications for fuel cycle facilities have used a conservative, deterministic approach. The Committee has letters on record pointing out problems with using this approach (see Appendix C). Some of the most important problems are that using very conservative assumptions can mask risk-significant items and most conservative analyses are not accompanied by a robust uncertainty analysis.

In at least one recent instance, DOE has used a deterministic dose assessment based on best estimates [DOE, 2005]. This, when accompanied by a robust sensitivity and uncertainty analysis, might be appropriate for less complex fuel cycle facilities. While a probabilistic analysis based on conservative data and models could be performed, there is no evident benefit to doing so, and the conservatism would render the accompanying uncertainty analysis meaningless.

- One-Step COL: 10 CFR Part 70 allows for a one-step licensing process which means that the design and process details necessary to review the adequacy of a recycle facility would not be available until relatively late in the licensing process. This approach is likely to be more efficient for the NRC and less burdensome to the applicant than the traditional two-step licensing process for facilities containing well-established processes and equipment and where there is a base of licensing experience (e.g., reactors, uranium fuel fabrication plants). However, SNF recycle facilities have the potential to involve equipment, chemicals, and processes that are unfamiliar to NRC staff and that could necessitate multiple requests for additional information from licensees and/or extensive interactions between NRC staff and the licensee after license submittal to identify and resolve potential licensing issues. The proposed Yucca Mountain repository is an example of an unfamiliar facility where a two-step licensing process has been adopted and extensive pre-license-application interactions have occurred.

- Accommodating the Potential Future Diversity of 10 CFR Part 70 License Applications: 10 CFR Part 70 is used to license many nuclear material processing facilities other than those for fuel recycle. Such facilities are typically much smaller, less costly, and less complex than anticipated SNF recycle facilities to the point that imposing requirements appropriate for recycle facilities could be unduly burdensome to some applicants.

- Risk-Informed, Performance-Based[28]: A risk-informed regulatory approach is one in which risk provides an important insight for licensing a facility but where other considerations, such as cost and environmental impacts, can be balanced against the required extent of risk reduction. The ALARA philosophy epitomizes a risk-informed approach. Risk-informed regulations and licensing approaches apply in a wide range of situations, and the opportunities for focusing scarce resources on the most risk-significant items in very complex facilities would indicate its appropriateness in this instance. It is prudent for regulations for licensing fuel recycle facilities to include provisions that allow the regulator to make exceptions on a case-by-case basis.

- A corollary factor to a regulation being risk informed is that it is performance based. That is, the criteria for granting a license are expressed in terms of the requirements the applicant must meet but not the means by which the applicant meets the requirement.

---

[28] The Commission defined risk-informed regulation in its white paper [NRC, 1998] "Risk-Informed and Performance-Based Regulation" as "a philosophy whereby risk insights are considered together with other factors to establish requirements that better focus licensee and regulatory attention on design and operational issues commensurate with their importance to public health and safety."

For example, a regulation that requires that a dose limit be met is performance based, but one that requires use of a specific technology is not.

- Programmatic Specificity of Changes to 10 CFR Part 70: The NRC staff paper presenting options for licensing SNF recycle facilities focused on the DOE GNEP and the facilities currently being proposed by DOE. The scope, functional requirements, size, and timing of these facilities are still evolving and likely to change in unknowable ways in response to factors such as technology development, budget considerations, stakeholder input, and broader U.S. and international decisions on nuclear and energy policy. It would be inefficient to initially develop program-specific regulations and then have to revisit the regulations in the future for the purpose of generalizing them.

9.2. Impacts on Related Regulations

In addition to establishing the approach(es) to use for the primary licensing regulations for fuel recycle facilities, it will be necessary to evaluate the impact that recycle facilities and operations may have on other regulations that may be invoked in the licensing framework or that may need to be developed. The following sections discuss various features of fuel recycle facilities and operations and how these features may impact regulations other than the primary regulation.

9.2.1. Potential Impacts of New Radioactive Product, Effluent, and Waste Materials

9.2.1.1. Identification of New Product, Effluent, and Waste Materials from SNF Recycle

Fuel recycle facilities using any of the UREX variants would produce new radioactive product, effluent, and waste materials for which the current NRC regulatory system may not be adequate. Examples of new materials include the following:

- Recovered uranium containing small amounts of contaminants such as TRU actinides (e.g., ^{237}Np), fission products (e.g., ^{99}Tc), ^{232}U, and ^{236}U.

- A gaseous effluent stream from the fuel reprocessing plant that initially contains most of the intermediate-to-long-lived volatile radionuclides such as ^{129}I, ^{85}Kr, ^{14}C, and ^{3}H in the fuel fed to the plant. Historically in the United States, most (about 99 percent) of the ^{129}I has been removed from the effluent stream and managed as a solid waste. At present, the La Hague plant and THORP capture the iodine by caustic scrubbing and release it to the sea, relying on the enormous amount of iodine in the sea to provide adequate isotopic dilution. Caustic scrubbing also captures some of the ^{14}C, which is released to the sea. The new Rokkasho-Mura reprocessing plant will capture the iodine on a solid sorbent, the disposition of which has not yet been decided. Radionuclides in the gaseous effluent other than those mentioned are being released to the atmosphere.

- Spent fuel metal hardware containing small amounts of residual spent fuel and potentially the dissolver solids and ^{99}Tc that has been melted to form a monolithic or compacted waste form.

- Wastes containing a mixture of recovered cesium and strontium including the intermediate-lived radioactive isotopes $^{135,\,137}$Cs and ^{90}Sr, plus very small amounts of their short-lived (^{137m}Ba and ^{90}Y) isotopes and amounts of their stable products (^{135,137}Ba and ^{90}Zr) that are eventually equivalent to the initial amounts of $^{135,\,137}$Cs and ^{90}Sr.

- Substantial volumes of materials and equipment containing greater than 100 nCi/g of TRU radionuclides.

- A fission product waste stream containing lanthanides and other fission products that is less radioactive and decays more quickly than the HLW stream produced or planned for in the past.

Table 19 presents the estimated volumes, masses, radioactivity, thermal power, and classification of wastes from the UREX+1a flowsheet. Production of these wastes would raise a number of issues which are discussed in the following paragraphs.

9.2.1.2. Classification of Wastes

Classification of the wastes is an important determinant of how they must be treated, stored, transported, and disposed of. Under current law and regulations, the classification of the various wastes would range from Class A LLW to HLW. However, many of these wastes and the proposed management approach associated with them were not anticipated when the current waste classification system evolved, so the appropriateness of the classification remains open to question. Specific questions regarding waste classification include the following:

- It must be decided whether the cesium/strontium waste will require a waste determination and DOE decision that it is waste incidental to reprocessing so that it will not require disposal in a deep geologic repository.

- The stable end point of cesium decay is stable isotopes of barium. A waste containing barium is considered to be characteristically hazardous by virtue of its toxicity if the leach rate of the barium in standardized tests exceeds a prescribed limit. As a consequence, leaching tests will have to be performed on the cesium/strontium waste form to ascertain whether leached barium concentrations are too high and, if so, it must be decided whether the waste will require further treatment before disposal or be managed as a mixed waste.

- Existing technology can reduce the TRU element and other radionuclide concentrations in any uranium product deemed to be a waste low enough to be considered Class A LLW. Waste containing ^{85}Kr and ^{135}Cs in any concentration would be Class A LLW under the present system because these radionuclides are not listed in the waste classification tables in 10 CFR Part 61. Such wastes were not contemplated when the waste classification tables in 10 CFR Part 61 were finalized, and the appropriateness of these classifications requires further evaluation.

9.2.1.3. Waste Forms

Determination of the requirements for waste forms and packaging for wastes such as the volatile radionuclides, ^{137}Cs, and ^{90}Sr is necessary to define how the waste must be treated. This determination also has a significant impact on the selection of recovery processes for some species such as those in gaseous effluents. Waste form options for the volatile radionuclides were studied in the 1970s and 1980s, but process development was not completed, and a preferred waste form was not selected. Selection of a waste form for ^{85}Kr is particularly challenging because it is a nonreactive gas under all but extreme conditions. Large amounts of

^{137}Cs and ^{190}Sr have been made into chloride and fluoride chemical forms, respectively, and stored by DOE in water pools at Hanford for decades. However, the chloride and fluoride do not appear to be appropriate forms for near-surface disposal such as that being suggested by DOE. DOE has proposed using an aluminosilicate waste form. 10 CFR Part 61 does not address waste forms or packaging for these materials even if they were to be classified as Class C or less. There is no regulation addressing the form of GTCC LLW.

9.2.1.4.   Distribution of Radionuclides in Product, Effluent, Waste, and Process Streams

There is no technical basis for predicting the distribution of some radionuclides in recycle plant product, effluent, waste, and process streams. This distribution is necessary to determine the process routing required by each stream (e.g., does a stream that contains iodine that would be released during subsequent high-temperature processing need to go to iodine recovery?). The NRC also needs to know this distribution to estimate doses from release of effluents or disposal of wastes and to evaluate the consequences of accidents. Important radionuclides questions in this regard include the following:

- Tritium: To what extent is the zirconium tritide on the cladding surface released during voloxidation, during acid dissolution of the SNF, and during the melting of the fuel assembly hardware to yield the waste form proposed by DOE?

- Iodine: Do iodine species from that are not trapped by available technologies and that might exceed the allowable release of about 0.5 percent form? What fraction of the iodine is associated with dissolver solids, and what fraction is released when the dissolver solids are included in the final waste form that involves high-temperature melting?

- Technetium and neptunium: What fraction of the technetium is associated with the dissolver solids? Of the neptunium and dissolved technetium, a small but potentially significant fraction can be found in various waste streams. What fractions are associated with the various waste streams and products from the reprocessing plant?

- Cladding: How much of the SNF remains with the cladding? Is the radionuclide distribution the same as in the SNF, or are some elements preferentially associated with the cladding? This is somewhat important in a waste disposal situation but would be very important if DOE proposals involving recycling the cladding material become reality.

9.2.1.5.   Disposal Technology

Requirements for disposal technologies appropriate for some of the wastes listed above have not been determined. For those wastes classified as GTCC, the technology and possibly a specific site may be identified as part of the ongoing DOE effort to prepare an EIS on this subject. The NRC will license the GTCC disposal facility using a regulatory framework that has not been decided. However, it is not evident that the EIS will consider potential GTCC wastes that are unique to recycle, such as cladding waste, possibly ^{137}Cs and ^{90}Sr (depending on when these are classified), miscellaneous wastes containing greater than 100 nCi/g TRU (e.g., equipment and analytical wastes, protective equipment, HEPA filters), and wastes containing ^{99}Tc, ^{129}I, and ^{14}C.

Identification of requirements for an appropriate disposal technology (i.e., the acceptability of near-surface disposal and conditions for same) for intermediate-lived radionuclides such as ^{85}Kr and tritium may depend on the ability of the selected waste form or package to contain substantial inventories and concentrations of these radionuclides until they decay to innocuous levels.

Uranium recovered from fuel reprocessing may exceed its demand, thus leading to the potential need to dispose of some of it. Determination of the acceptability of this uranium for near-surface disposal will need to consider the potential risks from species such as ^{237}Np and ^{99}Tc that are often more mobile than uranium under the geohydrological conditions that prevail near the surface at many sites and the effect of the $^{232,\,236}$U on the radiological impacts of the uranium. The NRC staff is undertaking an analysis of whether depleted uranium warrants inclusion in the waste classification tables in 10 CFR Part 61 pursuant to Commission direction [NRC, 2005].

9.2.1.6. Repository Licensing Regulations

Use of any of the UREX flowsheets for recycle would change the fundamental nature of a deep geologic repository to the point that the requirements in existing repository regulations would require reevaluation. Removing essentially all of the actinides (uranium and heavier), ^{137}Cs, ^{90}Sr, ^{99}Tc, and ^{129}I, and potentially the cladding, tritium, ^{14}C, and ^{85}Kr, from the repository would result in a compact repository waste that would generate considerably less penetrating radiation and decay heat that would decline much more quickly than in the case of SNF or traditional HLW. The amount of actinides and long-lived radionuclides that dominate risk estimates for the currently proposed repository would be reduced to levels that might cause other radionuclides that are presently not risk significant to become dominant in performance assessments. If some of the long-lived wastes mentioned above (technetium, iodine, carbon, cladding, and solid wastes containing some TRU elements) were to be disposed of in the deep geologic repository, the waste volume would increase somewhat, and the wastes would introduce some radionuclides important to public risk in new waste forms for which there is little experience in predicting long-term performance.

Consequently, aspects of existing regulations and guidance concerning repository licensing that are driven by decay heat, penetrating radiation, the actinides, the degradation rates of the spent fuel cladding and matrix, and the dominance of radionuclides such as ^{99}Tc and ^{237}Np may become irrelevant. On the other hand, the performance of multiple (and presently unknown) waste forms tailored to specific radioelements over very long time periods could become very important. The implications of this for the requirement to predict the performance of the repository to the time of peak dose have yet to be determined.

9.2.1.7. Uranium Handling and Disposal Facilities

The additional radionuclides in recovered uranium as compared to unirradiated uranium need to be considered when recycling the uranium to enrichment plants or handling it in other parts of the fuel cycle. The nonuranium isotopes tend to accumulate in certain portions of enrichment equipment and to be concentrated into a waste stream by decontamination operations during maintenance. This requires that enrichment plants have features to: (1) process wastes containing TRU and fission product elements and (2) detect beta-emitting radionuclides and distinguish among alpha-emitting radionuclides. The ^{236}U is a neutron absorber that detracts from the value of the recycled uranium and leads to increased production of ^{237}Np in fuel made from it. While present in very small quantities (about 1 part per billion by weight), decay of the ^{232}U in the recovered uranium to a ^{208}Tl decay product that emits a very penetrating 2.62 MeV gamma ray must be taken into account in the design of facilities for handling recycled uranium.

9.2.2. Novel Facilities

9.2.2.1. Cesium/Strontium Storage/Disposal Facility

Fuel recycles using a UREX or similar flowsheet would require facility types that have not been licensed in decades, if ever. Section 8 discussed regulatory issues concerning many of the major facilities, and Section 9.2.1 addressed issues in licensing a GTCC disposal facility, and those discussions will not be repeated here. However, there may be needed for a disposal technology not anticipated in existing regulations, specifically engineered near-surface interim storage facility that could store $^{135, 137}$Cs and ^{90}Sr waste forms for about 300 years, at which time the radionuclides will have decayed to less-than-Class-C levels. At that time, the storage facility could be converted to a disposal facility with the waste forms remaining in place. Use of this type of facility is one way to increase the capacity of the repository because it removes a major source of decay heat from the repository. This approach raises regulatory issues such as the following:

- whether the cesium/strontium waste is classified when it is produced or after the monitored interim storage period

- whether a near-surface facility containing radionuclides emitting considerable amounts of heat and penetrating radiation can be reliably designed, built, and maintained for as long as 300 years

- whether such a long-term storage facility would be suitable for conversion to a permanent disposal facility at that time and what technology would be used in such a conversion.

9.2.2.2. Storage Facility for Transuranic Element Product

Construction and operation of a fuel reprocessing plant before actinide burner reactors are available would result in the need to store significant quantities of TRU actinide products containing neptunium, plutonium, americium, and curium, possibly mixed with fission products emitting penetrating radiation to provide some degree of self-protection, until actinide burner reactors become available. Such a scenario would involve regulatory considerations of the acceptable form and technology for storing such a product and how best to safeguard it.

9.2.3. Novel Process Streams and Paradigms for Safeguards and Security

A fundamental feature of the UREX flowsheets approach is that fissile material (primarily plutonium) is never completely separated from other radionuclides. In particular, the UREX approach calls for the plutonium to remain mixed with other radionuclides (e.g., other actinides, possibly some fission products) that impart some degree of self-protecting characteristics by releasing penetrating radiation. It is axiomatic that any two substances can be separated with sufficient effort, although the magnitude of the effort can vary from trivial to impractical. Current levels defining what amount of radiation is "self-protecting"[29] (e.g., 100 R/h) were conceived with

---

[29] The term "self-protecting" is an arbitrary classification of protection derived from the radiation dose associated with irradiated spent commercial fuel. It is generally taken to be the protection afforded by a dose rate of 100 R/h, which is assumed to be high enough to deter the potential theft of the spent fuel or of anything else having at least that dose rate.

a spent fuel assembly (180 to 500 kilograms heavy metal (HM) of spent fuel) in mind. It is not clear that current dose values are applicable to or even achievable for amounts of plutonium and fission products on the order of 10 kilograms. The foregoing raises issues such as how much penetrating radiation from what amount of material is enough to be self-protecting, how difficult does the separation of plutonium from other radionuclides have to be for the plutonium to be deemed self-protecting, and how is the concept of a self-protecting material factored into the safeguards and security paradigms that will be used in the recycle facilities, if at all. Regulations that will be used to support licensing must consider these questions.

9.2.4. Evaluation of Integrated Plant Performance

The UREX flowsheets are extraordinarily complex. In essence, a UREX flowsheet includes at least four interconnected processes operating in series. Each of these processes is as complex as the traditional PUREX process, and some promise to be more difficult to control (e.g., TALSPEAK). The processes are also likely to include many types of equipment beyond those included in PUREX plants to recover additional radionuclides from gaseous effluents, to treat the many new waste streams mentioned previously, and to recycle various materials to reduce amounts of effluents and wastes. These complexities indicate that such a plant is likely to be difficult to operate, requiring extensive and expensive operator training and sophisticated control and monitoring systems. Of more relevance to a regulator are the difficulty and resource requirements of developing the technical capability (expertise and analytical tools) to evaluate whether such a complex system can be safely operated under normal and accident conditions, which involves predicting the behavior of myriad pieces of equipment, the piping connecting them, and the radioactive materials in them. This task is even more difficult because of factors such as (1) the potential for unexpected minor species to appear in a unit operation because of upsets in internal recycle which can cause unanticipated hazardous conditions, and (2) the ramifications of an equipment failure and quick shutdown of an entire interconnected plant. Interprocess surge capacity may be a very important design feature in the management of such problems.

9.2.5. Design and Operation with Decommissioning in Mind

The NRC Commissioners have stated that an important goal in licensing nuclear facilities in general and fuel recycle facilities in particular, is to include requirements to minimize historical problems in decommissioning the facilities at the end of their operating life. This is a relatively new NRC requirement, and one that is very worthwhile. The decommissioning process affects important issues such as residual site contamination, stored wastes, environmental problems, the health and safety of cleanup workers, and cost. In turn, the way in which facilities are designed and operated determines the manner in which decommissioning is performed. Thus, meeting requirements to facilitate ultimate facility decommissioning must be part of obtaining a license to construct and operate fuel recycle facilities. Specifying such requirements will be challenging because (1) the commercial plant designer and the ultimate plant operator will want freedom to build the plant in a way that efficiently accomplishes the principal plant mission (namely, spent fuel recycle), and (2) the experience on which to base the requirements for recycle facilities is not yet available.

The NRC [NUREG, 2007] has provided consolidated general decommissioning guidance, and the NRC and EPA have signed a memorandum of understanding on decommissioning [MOU, 2002]. Beyond this, the Committee and NRC staff is presently working within their respective mandates to gather lessons learned related to the decommissioning of fuel recycle facilities.

This information is planned for use as a basis for recommending additional requirements to be included in existing or new regulations concerning the design, construction, and operation of fuel recycle facilities to facilitate decommissioning and license termination.

9.3. NRC Test Facilities

As is evident from the foregoing, recycle facilities that are capable of meeting GNEP goals will involve many processes and pieces of equipment that have never been used on a commercial scale or in licensed facilities. Consequently, there is no established basis for assessing the performance and safety implications of these processes and equipment. It is expected that DOE will base its assessments on information it obtains from lab-scale tests using SNF in hot cells at its national laboratories, plus engineering or pilot-scale equipment testing possibly using uranium.

When licensing facilities, the NRC normally performs confirmatory research to validate key data and assumptions made by an applicant. In the case of recycle facilities, such research would require highly specialized facilities (e.g., hot cells) and equipment that is available in only a few places, none of which are part of the current NRC community. The lack of NRC infrastructure for SNF recycle raises the issue of how the NRC will perform confirmatory research. Options include observation of DOE experiments, contracting with DOE or possibly with the very few commercial firms for the use of hot cells, and collaboration with other countries to obtain access to hot cells.

9.4. Operator Licensing Examinations

It will be necessary to create and grade licensing examinations for fuel recycle facility operators at several levels of competence and responsibility. Facilities such as reprocessing plants require several levels of operator training. In addition, there is "cross training" in plant operations for other personnel such as guards and maintenance crews. Experience has shown that training and qualifying plant operators is difficult, time consuming, and expensive. Finding people qualified to prepare and administer proficiency examinations will be challenging. The elapsed time since such examinations were last administered and the likely requirement for having to add new examination topics, such as those related to proliferation prevention and detection and safeguards make this an important area for consideration.

9.5. Sigma Inventory Difference Requirements

Table 16 indicates the major differences among the IAEA, NRC, and DOE on the requirements for the permissible significant (sigma) plutonium IDs and the frequency of both long-term shutdown inventory and interim frequency requirements. This disparity could have a significant impact on facility design and must be addressed and resolved to the extent practicable for any recycle facility licensed in this country.

9.6. Timing and Urgency

As a practical matter, the number and timing of license applications for fuel recycle facilities are important factors in deciding the nature and urgency of the regulatory approach to be used. As this paper is written, the schedule announced by DOE for building recycle facilities extends no further than a major decision to be made on whether and how to proceed based on the contents of a PEIS now in preparation, although DOE has established a planning milestone for initial operation of an SNF reprocessing plant in 2020 [GNEP, 2007b].

9.6.1. Time Required To Prepare for Review of a License Application for a UREX Flowsheet

Assuming that DOE decides to develop, demonstrate, and deploy one of the UREX variants in a first-of-a-kind recycle facility a number of potential licensing issues will need to be addressed:

- Considerable work remains to be done in taking processes that have been tested on SNF separately only at a lab scale through a larger scale integrated demonstration. Also, equipment must be tested using nonradioactive materials or uranium. The SNF reprocessing demonstration and equipment testing can proceed in parallel.

- Considerable work is needed before the reprocessing plant off-gas system can be designed:

    - Integrated off-gas systems likely to be acceptable in the United States (i.e., no release of ^{129}I to the sea, ^{85}Kr recovery, potential recovery of ^{3}H and ^{14}C) have never been operated in any large facility.

    - Separate processes for the recovery of ^{85}Kr, ^{3}H, and ^{14}C the last three species have never been operated in any large-scale facility.

    - Development of processes for ^{85}Kr, ^{3}H, and ^{14}C was never completed, although some work on ^{85}Kr removal processes has continued.

    - Disposal destinations and waste forms are not yet established. Significant studies and development work will be required.

    - The process of establishing radionuclide release limits for reprocessing plant gaseous effluents must be reengaged because it was never completed.

    Development of release limits for radionuclides in reprocessing plant gaseous effluents and completion of the required technology development are likely to be on the critical path to a license application because of the need to develop an acceptable conceptual approach to establish the limits, develop cost estimates for various levels of radionuclide removal and risks associated with each level as a basis for the limits, go through the process to establish the limits, and undertake the necessary technology development and demonstration. These steps can be performed in parallel only to a limited extent.

- After the foregoing work is completed, a facility design, license application, and other environmental documents will require preparation.

- Decisions must be made on a number of policy issues (e.g., ISA versus PRA, performance-based requirements or not, how to license a complex facility without unduly burdening applicants for simpler facility licenses) before work to establish the primary licensing regulation can begin in earnest, and an analysis (already underway) to evaluate gaps in other regulations is needed.

- Modifying an existing regulation (or developing a new regulation) to be a risk-informed licensing regulation for a facility as complex as a reprocessing plant using a UREX flowsheet or equivalent is a major undertaking.

- The provisions of many regulations (other than the primary licensing regulation) identified in the gap analysis will require revision. Regulations where some degree of change is likely to be required include 10 CFR Parts 30, 50, 51, 52, 61, 63, 73, 74, the framework for civilian waste classification, and the regulation for licensing disposal of GTCC waste. This may include extensive involvement in developing the limits for radionuclide releases to the gaseous effluent.

- Most of the guidance concerning SNF reprocessing plants dates from the mid-1970s and will require revision to reflect current standards, technology, and regulations.

- Conducting all of the foregoing activities in parallel is likely to require a large "bubble" in expert staff levels and budgets that may not be available. If increased staff and budget are not available, an alternative approach would involve prioritizing the above activities and undertaking them more sequentially, which would increase the time required.

The preceding discussion and the uncertainties mentioned indicate that the time required for DOE to submit a license application for a UREX-based SNF reprocessing plant is commensurate with the time required for the NRC to develop the necessary suite of regulations and supporting guidance.

9.6.2. Time Required To Prepare For Review of a License Application for a Modified PUREX Flowsheet

The premise of the timing estimates in the preceding section is that DOE will propose to deploy a UREX flowsheet and the NRC will review a license application for the plant. However, DOE has recently indicated that the initial fuel reprocessing plant may be based on a PUREX flowsheet modified so that it does not produce a pure plutonium product. The implication is that the facility design would allow other capabilities (e.g., cesium/strontium removal, separation of a product composed of all TRU elements) to be incorporated in modules to be added in the future. This approach might involve storage of the PUREX raffinate as an acidic liquid pending addition of the new modules to process the stored raffinate.

This approach would have two important implications:

(1) Most of the technology required to prepare a license application exists. The difficult aspects of UREX (relatively new technologies needing integration with a modified PUREX process and each other) would be deferred pending additional development.

(2) The exception to the preceding item concerns release limits for radionuclides in gaseous effluents. As discussed earlier, the regulations providing the design basis for limiting such releases and the technology for meeting these limits are not yet available. Such limits and technology need to be established to reprocess SNF using any flowsheet including a modified PUREX.

Under a modified PUREX approach, it would be possible to prepare a facility design, license application, and supporting environmental documentation within about 5 years (allowing time for budgeting plus design and document preparation) with one important exception—the off-gas treatment system. As described earlier, designing the off-gas system depends first on establishing release limits for key radionuclides in the gaseous effluent and then developing an off-gas treatment system capable of meeting the limits. A scenario involving a modified PUREX approach still requires specification of release limits for radionuclides in the gaseous effluent but requires the limits even earlier than in a scenario where DOE would have to complete development and demonstration of a UREX flowsheet. Representatives of the two major reprocessing organizations stated in the Committee's July 2007 meeting [ACNW&M, 2007] that establishing release limits was high priority.

10. OTHER IMPORTANT ISSUES RELATED TO LICENSING

10.1.   Completion of Generic Environmental Documentation and Standards

In the 1970s when nuclear fuel recycle was being aggressively pursued by the AEC, ERDA, and DOE, efforts were undertaken to prepare a generic (programmatic) environmental impact statement (GEIS) on nuclear fuel recycle. EPA began to develop standards for radionuclide releases from recycle facilities. This effort was stimulated by and intertwined with the license application for BNFP. Some work continued on both fronts even after President Carter banned nuclear fuel reprocessing in the United States and the BNFP license application was withdrawn.

The GEIS and BNFP licensing efforts became the platform for a contentious debate over whether the United States should pursue fuel recycle. The GEIS effort ended with the publication of the GESMO document. That document did not encompass the recycle scenarios now being proposed and consequently is probably not useful. However, DOE has recently initiated preparation of what is essentially the follow-on to GESMO by issuing a Notice of Intent [DOE, 2007] to prepare a GNEP PEIS.

EPA initiated development of environmental radiation protection standards for the nuclear fuel cycle in the 1970s. Briefly, the approach used by EPA was to assess the ability of existing and developing sequences of processes for removing various radionuclides from effluent streams as expressed in terms of the collective dose reduction that would result from each incremental process. The Agency evaluated the cost of each incremental process using then-standard cost-benefit techniques. At some point, the cost per unit dose reduction ($/man-rem) from the last incremental process was deemed excessive, and the extent of radionuclide removal without the last incremental process became the bases for the standard. EPA performed cost-benefit analyses for all major steps of the nuclear fuel cycle (e.g., uranium mining and milling, reactor operation, and reprocessing) based on technical studies supported by EPA and the NRC. The result is codified in 40 CFR Part 190. Of particular relevance to fuel recycle is 40 CFR 190.10(b) which limits the release of ^{85}Kr and ^{129}I from normal operations of the uranium fuel cycle. Because fuel reprocessing is the only step of the nuclear fuel cycle that could release significant amounts of these radionuclides during normal operations, these limits are effectively release limits for the fuel reprocessing gaseous effluent. The NRC adopted this standard in 10 CFR Part 20.1301(e).

From the perspective of decades of hindsight, 40 CFR Part 190 raises the following concerns:

- The factors by which ^{85}Kr and ^{129}I must be reduced are approximately 7-fold and 200-fold, respectively. The evaluation that led to these factors was based on effluent control technologies that were under development at the time but had not been demonstrated or deployed. Because fuel recycle was abandoned, development was never completed. Thus, meeting the standard with available technologies may not be feasible.

- Background information accompanying the standard indicated that studies concerning limits on releases of ^{14}C and ^{3}H were underway. These studies remain uncompleted, and thus, the standard may be incomplete.

- The cost-benefit approach used in the analyses involved calculating the collective dose by integrating very small doses over very large populations and distances and comparing the collective dose to then-common metrics such as a limit of $1000/man-rem to determine whether additional effluent controls were justified. As Committee letters and the International Commission on Radiological Protection have observed, such a comparison is questionable and should not be used in favor of using dose to a maximally exposed individual or critical group.

- The scope of 40 CFR Part 190 does not include refabrication of fuels enriched with plutonium or actinides other than uranium. This addition would presumably be necessary for fuel recycle to proceed. The standard is therefore not complete.

In summary, the EPA standard on which effluent release limits are based may impose requirements that are infeasible in the near-term, is incomplete, and is based on analysis techniques that have become questionable over the years.

10.2. Obtaining Adequate Numbers of Qualified Staff

Implementing fuel recycle will require a substantial number of staff knowledgeable about the technical and regulatory aspects of fuel recycle facility design and operation. The design and operation of the fuel reprocessing and recycle fuel fabrication facilities are particularly challenging because staff members trained as nuclear chemical operators and engineers are required.

With the virtual disappearance of work in the civilian nuclear fuel cycle in the 1976–1985 timeframe and the cessation of defense reprocessing activities in the following decade, workers moved into other areas and most have now retired with their expertise not having been replaced because there has been little demand. While the Nuclear Navy continues to offer a good supply of reactor operators, there is no parallel source for nuclear chemical operators, who usually have an associate degree and are then trained on the job. As noted earlier, recycle facilities are very complex, and the failure rate of those examined can be high, as evidenced by the experience at NFS and BNFP for new recruits. Similarly, nuclear chemical engineers historically have had an undergraduate degree in chemical engineering and obtained graduate degrees in nuclear engineering and then practical experience on the job. Unfortunately, nuclear chemical engineering programs have been drastically reduced or eliminated, and many of the faculty that taught this subject have retired. Organizations performing fuel recycle R&D, designing and operating recycle facilities, and regulating recycle facilities will be seeking this same expertise, especially that of nuclear chemical engineers, thus exacerbating the supply-demand imbalance for this very limited expertise.

10.3. Potential International Issues

The goals of the GNEP include having once-through and recycle facilities in the United States providing services (fuel supply, fuel take-back) as a primary component. The relationship that must be established among the various countries for this to occur with confidence is not yet clear. However, with substantial amounts of U.S. fuel going to many other countries and then being returned to this country, a more focused interaction may be needed between the NRC and foreign regulators to ensure that U.S. fuels are acceptable internationally and that fuel irradiated in another country has an acceptable pedigree for its return.

10.4. Interface between NRC and DOE Regulatory Authorities

DOE regulates most of its activities under its own authority, while the NRC regulates licensees engaged in civilian and commercial nuclear activities. Decisions on whether a particular facility having significant DOE involvement or funding is regulated by DOE or the NRC, especially if it is a relatively unique facility, are often made on a case-by-case basis. In the case of the projected fuel recycle facilities, a patchwork of regulations could arise, with DOE regulating some facilities that interface with other NRC-regulated facilities (e.g., a fuel refabrication plant and the waste management facilities serving it). This could pose challenges concerning compatibility and consistency of regulatory requirements and evaluating safety, especially in cases where material moves between facilities. This scenario is occurring at the MOX fuel fabrication plant at SRS, but it could be far more complex for a reprocessing plant with its myriad wastes and recycle streams.

Even for activities regulated under DOE authority, the design and operation of the facilities provide an excellent opportunity to educate and train NRC staff for licensing subsequent facilities and to obtain insights useful in developing or modifying NRC regulations to license future commercial facilities. Of particular note is a stepwise, end-to-end demonstration of the UREX+1a flowsheet now underway at ORNL [Binder, 2007], which begins with SNF receipt and ends with fabrication of fuels containing TRU elements and use of waste materials (e.g., technetium, cesium/strontium) to develop treatment processes.

THIS PAGE IS INTENTIONALLY LEFT BLANK.

REFERENCES

General

Proceedings of the First United Nations International Conference on the Peaceful Uses of Atomic Energy, United Nations publication, Geneva, Switzerland, August 1955.

Proceedings of the Second United Nations International Conference on the Peaceful Uses of Nuclear Energy, United Nations publication, Geneva, Switzerland, 1958.

"Chemical Aspects of the Nuclear Fuel Cycle: A Special Issue," Radiochimica Acta, Vol. 25, No. 3/4, 1978.

"Recent advances in reprocessing of irradiated fuel: Nuclear Engineering part XX," Chemical Engineering Progress Symposium Series, 94, Vol. 65, 1969.

"Aqueous Reprocessing Chemistry for Irradiated Fuels," Brussels Symposium, OECD European Nuclear Energy Agency, 1963.

Nonproliferation Treaty, Article IV: Peaceful Uses of Nuclear Energy, Statement to the 2005 Review Conference of the Treaty on the Nonproliferation of Nuclear Weapons, Christopher Ford, Principal Deputy Assistant, Bureau of Verification and Compliance, New York, May 18, 2005.

"Status and Assessment Report on Actinide and Fission Product Partitioning and Transmutation," OECD Nuclear Energy Agency, May 1999.

Specific

ACNW&M (2002). ACNW&M letter from George M. Hornberger, Chairman ACRS, to Richard A. Meserve, Chairman NRC, Subject: Risk-Informed Activities in the Office of Nuclear Material Safety and Safeguards, ADAMS Accession No. ML072820458, January 14, 2002.

ACNW&M (2006). ACNW&M letter dated May 2, 2006, from Michael T. Ryan, Chairman ACNW&M, to Nils J. Diaz, Chairman NRC, Subject: Risk-Informed Decision-Making for Nuclear Materials and Wastes.

ACNW&M (2007). NRC/ACNW&M, May 2007 meeting transcript.

AIChE 1969. American Institute of Chemical Engineers, "Recent Advances in Reprocessing of Irradiated Fuel," W. A. Rodger and D. E. Ferguson, Eds., Nuclear Engineering part XX, No. 94, Vol. 65.

Albenesius (1983). E.L. Albenesius, "Tritium Waste Disposal Technology in the U.S.," presented at the DOE/CEA Meeting on Defense Waste Management, November 9, 1983, Knoxville, Tennessee, DP-MS-83-114 (CONF-8311105-2), January 1983.

ANL (1983). L.E. Trevorrow et al., "Compatibilities of Technologies with Regulations in the Waste Management of H-3, I-129, C-14, and Kr-85, Part II, Analysis," ANL-83-57, Part II, November 1983.

ANL (2002). Argonne National Laboratory, *Frontiers 2002 Research Highlights*, http://www.anl.gov/Media_Center/Frontiers/2002/index.html.

AREVA (2007a). AREVA NC, "Briefing to the NRC Advisory Committee on Nuclear Waste," May 15, 2007.

AREVA (2007b). AREVA NC, "Monitoring results," tables updated monthly on the AREVA Web site http:www.cogemalahague.com.

ASTM (2007). American Society for Testing and Materials, "Standard Specifications for Nuclear-Grade Uranyl Nitrate Solutions or Crystals," C 288-03.

Ayer (1988). J.E. Ayer et al, "Nuclear Fuel Cycle Facility Accident Analysis Handbook, " NUREG-1320, May 1988.

Azizova (2005). T.V. Azizova, N.G. Semenikhina, and M.B. Druzhinina, "Multi-organ Involvement and Failure in Selected Accident Cases with Acute Radiation Syndrome Observed at the Mayak Nuclear Facility," British Journal of Radiology, Supplement 27, 30–35, British Institute of Radiology, 2005.

Barre (2000). B. Barre and H. Masson, "State of the Art in Nuclear Fuel Reprocessing," Safe Waste 2000, October 2-4, 2000

Benedict (1981). M. Benedict, T. Pigford, and H. Levi, "Nuclear Chemical Engineering," Second Edition, McGraw-Hill.

Bibler (2002). N.E. Bibler and T.L. Fillinger, "DWPF Vitrification—Characterization of the Radioactive Glass Being Produced During Immobilization of the Second Batch of HLW Sludge," WSRC-MS-2000-00533.

Binder (2007). J.L. Binder, "The GNEP Spent Fuel Recycle Coupled End-to-End Demonstration," presentation to the National Academy of Sciences, July 17, 2007.

Blaga (1985). A. Blaga and J. J. Beaudoin, "Polymer Concrete," Canadian Building Digest, CBD-242, November 1985.

Blomeke (1972). J.O. Blomeke and J. J. Perona, "Storage, Shipment, and Disposal of Spent Fuel Cladding," ORNL/TM-3650, January 1972.

Boullis (2006). B. Bouliss, P. Barron, and B. Lorrain, "Progress in Partitioning: Activities in ALANTE," 9th Information Exchange Meeting on Partitioning and Transmutation, OECD Nuclear Energy Agency, September 25–29, 2006.

Bouchard (2005). Jacques Bouchard, "The Closed Fuel Cycle and Non-Proliferation Issues," Global 2005, Tsukuba, October 11, 2005.

Cazalet (2006). J. Cazalet, CEA, "Future Nuclear Cycle Systems and Technology Enhancing Proliferation Resistance," presentation in Panel 3 of the Nuclear Non-Proliferation Science and Technology Forum, Tokyo, Japan, May 18–19, 2006.

CCD-PEG (2003). J.D. Law, R.S. Herbst, D.R. Peterman, R.D. Tillotson, and T.A. Todd, "Development of a Cobalt Dicarbollide/Polyethylene Glycol Solvent Extraction Process for Separation of Cesium and Strontium to Support Advanced Aqueous Reprocessing," Proceedings from Global 2003, November 2003.

CCD-PEG (2006). J.D. Law, et al., "Development of Cesium and Strontium Separation and Immobilization Technologies in Support of an Advanced Nuclear Fuel Cycle," WM '06 Conference, Tucson, Arizona, February 26–March 2, 2006.

CEA (2007). O. Bernard-Mozziconacci, F. Devisme, J-L Merignier, and J. Belloni, "Colloidal silver iodide characterization within the framework of nuclear spent fuel dissolution," corresponding author email address: fredric.devisme@cea.fr.

Chandler (1956). J. M. Chandler and W. H. Lewis, "Metal Recovery Plant Activities During FY 1956," ORNL-2335.

China (1996). "Japan's Plutonium Programs: A Proliferation Threat?" Nonproliferation Review, page 8, Winter 1996.

Choppin (1987). G.R. Choppin, "Carbon-14 in the Palo Duro basin repository," BMI/ONWI/C-10, January 1, 1987.

Congress (2005). U.S. Congress, *Making Appropriations for Energy and Water Development for the Fiscal Year Ending September 30, 2006, and for Other Purposes*, Report 109-275 to accompany H.R. 2419 (subsequently P.L. 109-103, 119 STAT 2247, November 19, 2005).

Corliss (1954). W.R. Corliss and D.G. Harvey, "Radioisotopic Power Generation," Prentice-Hall.

Croff (1976). A.G. Croff, "Evaluation of Options Relative to the Fixation and Disposal of ^{14}C-Contaminated CO_2 as $CaCO_3$," ORNL/TM-5171, April 1976.

Croff (1978). A.G. Croff, et al., "Revised Uranium-Plutonium Cycle PWR and BWR Models for the ORIGEN Computer Code," ORNL/TM-6051, September 1978.

Croff (1980). A.G. Croff, "ORIGEN2—A Revised and Updated Version of the Oak Ridge Isotope Generation and Depletion Code," ORNL-5621, July 1980.

Croff (1982)). A.G. Croff, M.S. Liberman, and G.W. Morrison, "Graphical and Tabular Summaries of Decay Characteristics for Once-Through PWR, LMFBR, and FFTF Fuel Cycle Materials," ORNL/TM-8061, January 1982.

Croff (2000). A.G. Croff, et al., "Assessment of Preferred Depleted Uranium Disposal Forms," ORNL/TM-2000-161, June 2000.

Davidson (2007). D. Davidson, AREVA NC, Inc., email to J.H. Flack, NRC, Subject: "Additional Info on AREVA Reprocessing and Refabrication Operations," August 1, 2007.

Del Cul (2002). G.D. Del Cul, B.B. Spencer, C.W. Forsberg, E.D. Collins, and W.S. Rickman, "TRISO-Coated Fuel Processing to Support High-Temperature Gas-Cooled Reactors," ORNL/TM-2002/156, September 2002.

DHEC (2000). South Carolina Department of Health and Environmental Control, License No. 97, Amendment No. 48 for the Barnwell Waste Management Facility.

DNFSB (2001). Defense Nuclear Facilities Safety Board, "Chemical Process Safety, HB-Line Phase II," Staff Issue Report, Memorandum for J.K. Fortenberry.

DNFSB (2003). R.N. Robinson, D.M. Gutowski, W. Yeniscavitch, with assistance from J. Contardi, R. Daniels, and T.L. Hunt, "Control of Red Oil Explosions in Defense Nuclear Facilities," Defense Nuclear Facilities Safety Board, DNFSB/TECH-33, November 2003.

DOE (1986). U.S. Department of Energy, "Technology for Commercial Radioactive Waste Management," DOE/ET-0028, May 1986.

DOE (1998). U.S. Department of Energy, "TRUEX/SREX Demonstration" Innovative Technology Summary Report, DOE/EM-0419, December 1998.

DOE (2005). U.S. Department of Energy, "Initial Single-Shell Tank System Performance Assessment for the Hanford Site," DOE/ORP-2005-01.

DOE (2006a). U.S. Department of Energy, "Report to Congress: Spent Nuclear Fuel Recycling Program Plan," May 2006.

DOE (2006b). U.S. Department of Energy, "Notice of Request for Expressions of Interest in a Consolidated Fuel Treatment Center to Support the Global Nuclear Energy Partnership," 71 FR 151 44676–44679, August 7, 2006.

DOE (2007). U.S. Department of Energy, "Notice of Intent to Prepare a Programmatic Environmental Impact Statement for the Global Nuclear Energy Partnership," 72 FR 331, January 4, 2007.

EIA (2004). U.S. Energy Information Agency, "U.S. Spent Nuclear Fuel Data as of December 31, 2002," www.eia.doe.gov, October 2004.

EPA (1977). U.S. Environmental Protection Agency, "Environmental Radiation Requirements for Normal Operations in the Uranium Fuel Cycle," Title 40 Part 190 of *The Code of Federal Regulations.*

EPRI (2003). Electric Power Research Institute, "Scientific and Technical Priorities at Yucca Mountain," EPRI Report 1003335, December 2003.

Ford (1976). Presidential Documents, Vol 12, No. 44, p. 1626-1627, 1976.

Forsberg (1985). C. W. Forsberg et al, "Flowsheet and Source Terms for Radioactive Waste Projections," ORNL/TM-8462.

Ganguly (1997). C. Ganguly, "Sol-gel microsphere pelletization process for fabrication of $(U,Pu)O_2$, $(U,Pu)C$ and $(U,Pu)N$ Fuel pellets for the prototype fast breeder reactor in India," *Journal of Sol-Gel Science and Technology*, Vol. 9, No. 3, pp. 285-294.

GE-H (2007). General Electric-Hitachi. Presentation entitled "Commercializing Nuclear Pyroprocessing Technology" by E.P. Loewen and E.F. Saito to the ACNW&M, October 16, 2007.

GNEP (2007a). Office of Fuel Cycle Management, *Global Nuclear Energy Partnership Strategic Plan,* GNEP-167312, January 2007.

GNEP (2007b). Global Nuclear Energy Partnership, Technology Integration Office, *Global Nuclear Energy Partnership Technology Development Plan*, GNEP-TECH-TR-PP-2007-00020, July 25, 2007.

Goode (1973a). J.D. Goode, "Voloxidation: Removal of Volatile Fission Products from Spent LMFBR Fuels," ORNL/TM-3723.

Goode (1973b). J.D. Goode (Compiler), "Voloxidation—Removal of Volatile Fission Products from Spent LMFBR Fuels," ORNL/TM-3723, May 1973.

Hore-Lacey (2007). Ian Hore-Lacey, "Decommissioning nuclear facilities," *The Encyclopedia of Earth*, January 30, 2007.

IAEA (1980). International Atomic Energy Agency, "Separation, Storage, and Disposal of Krypton-85," Technical Report Series No. 199.

IAEA (1987). International Atomic Energy Agency, "Treatment, Conditioning, and Disposal of Iodine-129," Technical Report Series No. 276.

IAEA (2000). International Atomic Energy Agency. "Compilation and Evaluation of Fission Yield Nuclear Data," IAEA-TECDOC-1168, December 2000.

IAEA (2001). International Atomic Energy Agency, "The International Nuclear Event Scale (INES) User's Manual," IAEA-INES-2001, February 2001.

IAEA (2003a). International Atomic Energy Agency, "Status and Advances in MOX Fuel Technology," Technical Report Series No. 415, IAEA, 2003.

IAEA (2003b). International Atomic Energy Agency, "Guidance for the Evaluation of Innovative Nuclear Reactors and Fuel Cycles," IAEA-TECDOC-1362, June 2003.

IAEA (2004). International Atomic Energy Agency, "Management of Waste Containing Tritium and Carbon-14," Technical Report Series No. 421.

IAEA (2005). International Atomic Energy Agency. "Partitioning and Transmutation: Radioactive Waste Management Option," Workshop on Technology and Applications of Accelerator Driven Systems (ADS), ICTP Trieste, Italy, October 17-28, 2005.

IAEA (2006). International Atomic Energy Agency, "Characterization, Treatment and Conditioning of Radioactive Graphite from Decommissioning of Nuclear Reactors," IAEA-TECDOC-1521, September 2006.

IAEA (2007a). International Atomic Energy Agency, "Management of Reprocessed Uranium: Current Status and Future Prospects," IAEA-TECDOC-1529, February 2007 (see Appendix V for reprocessed UO_3 specifications).

IAEA (2007b). International Atomic Energy Agency, press release, "Improving the INES Scale," http://www.iaea.org/NewsCenter/News/2007/ines.html, September 20, 2007.

IAEA (2007c). International Atomic Energy Agency, "Communication received from the resident representative of the Russian Federation to the IAEA on the establishment, structure and operation of the International Uranium Enrichment Center," INFCIRC/708. June 8, 2007.

INFCE (1980). International Nuclear Fuel Cycle Evaluation, "Reprocessing, Plutonium Handling, Recycle (sic): Report of Working Group 4," INFCE/PC/2/4.

INPRO (2006a). International Project on Innovative Nuclear Reactors and Fuel Cycles, "Infrastructure Oriented Activities of the IAEA INPRO Project as a Response to the Opportunities and Challenges of Nuclear Energy," V. Sckolov et. al. , American Nuclear Society: 2006 Annual Meeting, June 4-8, 2006.

INPRO (2006b). "Methodology for Assessment of Innovative Nuclear Energy Systems as based on a defined set of Basic Principles, User Requirements and Criteria in the areas of Economics, Sustainability and Environment, Safety, Waste Management, Proliferation Resistance and recommendations on Cross Cutting Issues."

ISIS (2007). David Albright, "Shipments of Weapons-Usable Plutonium in the Commercial Nuclear Industry," The Institute for Science and International Security (ISIS), January 3, 2007.

Jantzen (2002). C.M. Jantzen, "Engineering Study of the Hanford Low-Activity Waste Steam Reforming Process," WSRC-TR-2002-00317, July 2002.

Jantzen (2004). C.M. Jantzen, A.D. Cozzi, and N.E. Bibler, "High-Level Waste Processing Experience with Increased Waste Loadings," WSC-TR-2004-00286.

Kee (1976). C.W. Kee, A.G. Croff, and J.O. Blomeke, "Updated Projections of Radioactive Waste to Be Generated by the U.S Power Industry," ORNL/TM-5427, December 1976.

Kessler (2006). J.H. Kessler, "Preliminary Analysis of the Maximum Disposal Capacity for CNSF in a Yucca Mountain Repository," presentation to the ACNW&M, April 19, 2006.

Kim (2006). J.G. Kim, "Glass-Bonded Zeolite Waste Form for Waste LiCl Salt," 15th Pacific Basin Nuclear Conference, Sydney, Australia, October 15–20, 2006.

Kitamura (1999). Motoya Kitamura, "Japan's Plutonium Program: A Proliferation Threat?" *Non-Proliferation Review*, Winter 1999, p. 8.

Kleykamp (1984). H. Kleykamp and R. Pejsa, "X-Ray Diffraction Studies on Irradiated Nuclear Fuels," Journal of Nuclear Materials 124 56-3.

Kouts (2007). C. Kouts, "Transportation, Aging, and Disposal (TAD) Canister System Status and Total System Model (TSM)," presentation to the ACNW&M, June 19, 2007.

Kursunoglu (2000). Behram N. Kursunoglu, Stephen L. Mintz, and Arnold Perlmutter (eds.), *The Challenges to Nuclear Power in the Twenty-First Century*, Kluwer Academic/Plenum Publishers, New York, 2000.

Laidler (2006). James Laidler, "The Global Nuclear Energy Partnership: Advanced Separations Technology Development," presentation to the ACNW, ADAMS Accession No. ML062090258, July 20, 2006.

Loewen (2007). E.P. Loewen, "Advanced Reactors: NUREG-1368 Applicability to Global Nuclear Energy Partnership," presentation at the 19th Annual Regulatory Information Conference, March 13–15, 2007.

Long (1978). J. T. Long, "Engineering for Nuclear Fuel Reprocessing," American Nuclear Society.

McGrail (2003). B.P. McGrail, et al., "Initial Suitability Evaluation of Steam-Reformed Low-Activity Waste for Direct Land Disposal," PNWD-3288, January 2003.

Mineo (2002). H. Mineo et al, "Study on Gaseous Effluent Treatment for Dissolution Step Nuclear Fuel Reprocessing," WM'02 Conference, February 24-28, 2002.

MOU (2002). MOU letter signed by Christine T. Whitman, U.S. EPA Administrator, and Richard A. Meserve, Chairman, U.S. NRC, September–October 2002.

NAS (2000). National Academy of Sciences, *Electrometallurgical Techniques for DOE Spent Fuel Treatment: Final Report*.

NEA (1999). Nuclear Energy Agency, Organization for Economic Cooperation and Development, "Evaluation of Speciation Technology," Workshop Proceedings, Tokai-Mura, Ibaraki, Japan, October 26–28, 1999.

NEA (2004). Nuclear Energy Agency, Organization for Economic Cooperation and Development, *Pyrochemical Separations in Nuclear Applications: A Status Report*, ISBN 92-64-02071-3, NEA Report No. 5427.

NEA (2006). Nuclear Energy Agency, Organization for Economic Cooperation and Development, "Advanced Nuclear Fuel Cycles and Radioactive Waste Management," NEA Report No. 5990.

NEI (2007). Nuclear Energy Institute, Web site at http:www.nei.org/resourcesandstats/nuclearstatistics/usnuclearpowerplants/.
NEP (2001). "National Energy Policy: Report of the National Energy Policy Development Group," May 16, 2001.

NERAC (2002). DOE Nuclear Energy Research Advisory Committee, "A Technology Roadmap for Generation IV Nuclear Energy Systems," Generation IV International Forum, GIF-002-00, December 2002.

NNI (1997). NNI—No Nukes Infosource, a publication of the Austrian Institute for Applied Ecology, March 11, 1997.

NRC (1976). U.S. Nuclear Regulatory Commission, "Final Generic Environmental Impact Statement on the Use of Recycle Plutonium in Mixed-Oxide Fuel in Light-Water Cooled Reactors: Health, Safety, and Environment," NUREG-0002, August 1976.

NRC (1981). U.S. Nuclear Regulatory Commission, "Draft Environmental Impact Statement on 10 CFR Part 61, 'Licensing Requirements for Land Disposal of Radioactive Waste,'" NUREG-0782.

NRC (1995). U.S. Nuclear Regulatory Commission, "Use of Probabilistic Risk Assessment Methods in Nuclear Regulatory Activities; Final Policy Statement," 60 FR 42622, August 16, 1995.

NRC (1998). U.S. Nuclear Regulatory Commission, "White Paper on Risk-Informed and Performance-Based Regulations, SECY-98-144, June 22, 1998.

NRC (2005). Memorandum and Order from A.L. Vietti-Cook, U.S. Nuclear Regulatory Commission, in the matter of Louisiana Energy Services, L.P. (National Enrichment Facility), CLI-05-20, October 19, 2005.

NRC (2006a). Memorandum from A.L. Vietti-Cook, U.S. Nuclear Regulatory Commission, to M.T. Ryan, Advisory Committee on Nuclear Waste, "Staff Requirements—COMSECY-05-0064—Fiscal Year 2006 and 2007 Action Plan for the ACNW&M," February 7, 2006.

NRC (2006b). Memorandum from A.L. Vietti-Cook, U.S. Nuclear Regulatory Commission, to M.T. Ryan, Advisory Committee on Nuclear Waste, "Staff Requirements—SECY-06-0066—Regulatory and Resource Implications of a Department of Energy Spent Nuclear Fuel Recycling Program," May 16, 2006.

NRC (2006c). U.S. Nuclear Regulatory Commission, "COMSECY-05-0064 SRM, Fiscal Year 2006 and 2007 Action Plan," February 7, 2006.

NRC (2007a). U.S. Nuclear Regulatory Commission, "Regulatory Options for Licensing Facilities Associated with the Global Nuclear Energy Partnership," SECY-07-0081, May 15, 2007.

NRC (2007b). U.S. Nuclear Regulatory Commission, "Staff Requirements—SECY-07-0081—Regulatory Options for Licensing Facilities Associated with the Global Nuclear Energy Partnership (GNEP)," SRM-SP07-0081, June 27, 2007.

NUREG (2007). U.S. Nuclear Regulatory Commission, "Consolidated Decommissioning Guidance," NUREG-1757, Vol. 1, Rev. 2; Vol. 2, Rev. 1; and Vol. 3, February 3, 2007.

NWPA (1996). Nuclear Waste Policy Act, July 31, 1996.

ORNL (2007). E.D. Collins, G.D. Del Cul, J.P. Renier, and B.B. Spencer, "Preliminary Multicycle Transuranic Actinide Partitioning—Transmutation Studies," ORNL/TM-2007/24, February 2007.

OSHA (2007). Occupational Safety and Health Administration. Web site at
http://www.osha.gov/SLTC/healthguidelines/zirconiumandcompounds/recognition.html

Pahl (1990). R.G. Pahl, D.L. Porter, C.E. Lahm, and G.L. Hoffman, "Experimental studies of
U-Pu-Zr fast reactor fuel pins in the experimental breeder reactor-II," Metallurgical and Materials
Transactions A, Vol. 21, Number 7, pp. 1863-1870, July 1990.

Pasamehmetoglu (2006). K.O. Pasamehmetoglu, AFCI Fuels Development National Technical
Director, Idaho National Laboratory, presentation to the ACNW&M, July 20, 2006.

PCA (2007). Portland Cement Association. Answer to frequently asked question "What are the
unit weights (densities) of cement and concrete" at
http://www.cement.org/tech/faq_unit_weights.asp.

Periera (2005). C. Periera et al., Demonstration of the UREX+2 Process Using Spent Fuel,"
WM'05 Symposium, February 27-March3, 2005.

Periera (2007). C. Periera et al., Demonstration of the UREX+1a Process Using Spent Fuel,"
WM'07 Symposium, February 25-March 1, 2007.

Phillips (2007). C. Phillips, EnergySolutions, email communication to the ACNW&M consultants,
July 25, 2007.

Roddy (1986). J.W. Roddy et al., "Physical and Decay Characteristics of Commercial LWR Spent
Fuel," ORNL/TM-9591/V1&R1, January 1986.

Schneider (2001). M. Schneider, X. Coeytaux, Y.B. Faid, Y. Marigac, E. Rouy, G. Thompson,
I. Fairlie, D. Lowry, D. Sumner, "Possible Toxic Effects from the Nuclear Processing Plants at
Sellafield at Cap de La Hague," EP/IV/A/STOA/, European Parliament, Directorate General for
Research, STOA Program, November 2001.

Sellafield (2005). British Nuclear Group, "Sellafield—Decommissioning & Termination Category
Summary, 35.13, 2005. Near Term Work Plan FY 2005/06 to 2007/08.

Simpson (2007). M. Simpson and I.T. Kim, "Separation of Fission Products from Molten LiCl-KCl
Salt Used for Electro-refining of Metal Fuels," INERI Project 2006-002-K, DOE/NE-0131.

TALSPEAK (1964). B. Weaver and F.A. Kappelmann, "A New Method of Separating Americium
and Curium from the Lanthanides by Extraction from an Aqueous Solution of an Aminopolyacetic
Acid Complex with a Monoacidic Organophosphate or Phosphonate," ORNL-3559, August 1964.

TALSPEAK (1999). M.P. Jensen and K.L. Nash, "Solvent Extraction Separations of Trivalent
Lanthanide and Actinide Ions Using an Aqueous Aminomethanediphosphonic Acid," Proceedings
of ISEC 1999, International Solvent Extraction Conference, July 11–16, 1999.

THORP (1984). J. Garraway, "The Behavior of Technetium in a Nuclear Fuel Reprocessing
Plant," I. Chem. E. Symposium Series No. 88, Extraction 84 Conference, Dounreay, Scotland.

THORP (1990a). I.S. Denniss and C. Phillips, "The Development of a Three Cycle Chemical Flowsheet to Reprocess Oxide Fuel," Paper 04-02, Proc. International Solvent Extraction Conference (ISEC), Kyoto, Japan, July 16–21, 1990.

THORP (1990b). I.S. Denniss and C. Phillips, "The Development of a Flowsheet to Separate Uranium and Plutonium in Irradiated Oxide Fuel," Proc. International Conference Extraction 90, Dounreay, Scotland, September 10–14, 1990 (published as I. Chem. E. Symposium Series No. 119, pp. 187–197, 1990).

THORP (1992). C. Phillips, "Uranium-Plutonium Separation by Pulsed Column in the First Cycle of the Three Cycle Thermal Oxide Reprocessing Plant," Waste Management 92, International Conference, University of Arizona, Tucson, March 1–5, 1992.

THORP (1993). C. Phillips, "Development and Design of the Thermal Oxide Reprocessing Plant at Sellafield," Trans I. Chem. E., Vol. 71, Part A, March 1993.

THORP (1999a). C. Phillips, "The Thermal Oxide Reprocessing Plant at Sellafield: Four Years of Active Operation of the Solvent Extraction Plant," International Solvent Extraction Conference (ISEC) 1999, University of Bellaterra (Barcelona), Spain, July 11–16 1999.

THORP (1999b). C. Phillips, "The Thermal Oxide Reprocessing Plant at Sellafield: Four Years of Successful Treatment of Irradiated Nuclear Fuel," Waste Management 1999, International Conference, University of Arizona, Tucson, February 28–March 5, 1999.

THORP (2000). Chris Phillips and Andrew Milliken, "Reprocessing as a Waste Management and Fuel Recycling Option: Experience at Sellafield," Waste Management 2000, University of Arizona, Tucson, February 27–March 2, 2000.

THORP (2006). Chris Burrows, Chris Phillips, and Andrew Milliken, "The Thermal Oxide Reprocessing Plant at Sellafield—Lessons Learned from 10 Years of Hot Operations and Their Applicability to the DOE Environmental Management Program," Waste Management 2006 Conference, Tucson, Arizona, February 26–March 2, 2006.

TRUEX (1998). Innovative Technology Summary Report, "TRUEX/SREX Demonstration," DOE/EM-0419, December 1998.

Vernaz (2006). Dr. Etienne Y. Vernaz, "Nuclear Waste in France: Current and Future Practice," Presented in the RSC Environment, Sustainability, and Energy Forum, Materials for Nuclear Waste Management, January 18, 2006.

Vondra (1977a). B. L. Vondra, "LWR Fuel Reprocessing and Recycle Program Quarterly Report for Period April 1 to June 30, 1977, ORNL-/TM-5987.

Vondra (1977b). B. L. Vondra, "Alternate Fuel Cycle Technologies Program Quarterly Report for Period July 1 to September 30, 1977," ORNL/TM-6076.

Weast (1968). *Handbook of Chemistry and Physics*, 49th, edition R.C. Weast, Ed., CRC Press.

West Valley (1981). Comprehensive information on the operations at the West Valley Plant, "Review of the Operating History of the Nuclear Fuel Service, Inc., West Valley, New York Irradiated Fuel Processing Plant," ORNL/Sub-81/31066/1, September 1981.

Wickham (1999). A.J. Wickham, G.B. Neighbour, and M. Duboug, "The Uncertain Future for Nuclear Graphite Disposal: Crisis or Opportunity?" IAEA Technical Committee Meeting on "Nuclear Graphite Waste Management," TCM—Manchester 99, pp. 13–28, October 18–20, 1999.

Wigeland (2007). R. Wigeland, Idaho National Laboratory, "Global Nuclear Energy Partnership: Waste Streams and Disposition Options," presentation to the National Academy of Sciences, April 4, 2007.

WNA (2006). World Nuclear Symposium, Annual Symposium 2006, remarks attributed to Sergei Kireinko by R.I.A. Novosti, London, England, September 7, 2006.

Wymer (1981). *Light Water Reactor Nuclear Fuel Cycle,* R. G. Wymer and B. L. Vondra, Eds. CRC Press.

GLOSSARY

A

aerosol—a suspension of fine particles in a gas

actinides—chemical elements with atomic numbers between 90 and 103

assay and accountability—analysis of a material and formally accounting for amounts of it

automation technology—technology perform tasks with reduced amounts of human intervention

B

batch dissolver—equipment used to dissolve material in batches rather than continuously

becquerel—one radioactive disintegration per second

biosphere—the surface region of the earth in which life can exist

bismuth phosphate process—separation process to recover plutonium from irradiated fuel by precipitating it using bismuth phosphate

blanket—regions surrounding the critical fissile core of a reactor for capturing neutrons in fertile material

blanket element—a unit (fuel rod) in the reactor blanket

boiling water reactor—a reactor in which the coolant water is permitted to boil

borosilicate glass—a type of glass containing oxides of the elements boron and silicon in addition to other glass formers

breeding ratio—the ratio of the number of fissile atoms produced to the number of fissile atoms consumed by a nuclear reactor

C

centrifugal contactor—a separation device in which two liquids are mixed in a rotating cylinder and then separated

centrifuge—as used in this paper, a device in which solids are separated from liquid by rapid rotation

ceramic—a hard, nonmetallic, inorganic material

chemical conversion process—a process in which material is converted from one chemical form to another

cladding hulls—pieces of the outer protective coating of nuclear fuel

climatic inversions—a weather condition in which the normal atmospheric layers are reversed in position

complex—as used in this paper, a chemical combination of two or more chemical species joined electrostatically to form a stable chemical entity

control rod—a rod containing isotopes of a neutron-capturing element used to control reactor reactivity

criticality—a condition wherein sufficient fissile material is present to sustain a nuclear chain reaction

crud—an undesirable solid material of undefined composition

D

decay heat—heat produced by the radioactive decay of radioisotopes

decontamination—the process whereby contaminants are removed from a material

denitrator—a piece of equipment in which a chemical nitrate is decomposed

deterministic—a regulatory approach to establishing goals that is based on analysis of what can go wrong and the consequences but not the probability of such problems

double-shell tank—a radioactive waste tank comprising a container within a container

E

effluents—material entering the environment from process equipment or a facility

electrochemistry—the relation of electricity to chemical changes using interconversion of chemical and electrical energy

electrometallurgical process—a process using electricity to produce metal

electrorefiner—a piece of equipment that uses electricity to separate and purify metals

environmental impact statement—a legally required document that presents and discusses the environmental and other effects of alternatives to building, modifying, or otherwise altering an existing facility or situation

equivalent enrichment—fissile characteristics of material calculated as though it behaved as if it were all enriched uranium

exothermic reaction—a chemical reaction that produces heat

extractable—the capability of being removed, typically from a liquid or gas stream

F

fast reactor—a nuclear reactor that does not substantially slow down the fission neutrons

fissile material—a substance that can undergo nuclear fission

fission products—elements produced when a nuclear material fissions

flowsheet—a diagram that shows the step-by-step movement of mass or energy using lines and conventional symbols

fuel assembly—a collection of fuel rods in a well-defined arrangement

G

gas-cooled fast reactor—a fast reactor in which the coolant is a gas such as helium

glove box—an enclosed container not having substantial radiation shielding in which an operator can handle hazardous material using attached gloves

graphite—a soft, solid, lustrous form of carbon that conducts electricity

ground water—water that travels through the earth below its surface

H

head end—the first steps in fuel reprocessing preceding solvent extraction, including fuel shearing and dissolving

heat exchanger—a device in which heat is transferred from one fluid to another without mixing the fluids

high-temperature gas-cooled reactor—a reactor capable of operating at high temperatures that is cooled with a gas and has a core and moderator made of graphite

hydrocyclone—a device in which a fluid is caused to rotate for the separation of the solid material it contains

I

isotope—one of the two or more atoms of an element having the same atomic number but different mass numbers

K

kernel—the essential central part of a substance (e.g., the fuel-containing portion of a microsphere)

L

lanthanide element—an element whose atomic number is greater than 56 and less than 72

lead-cooled fast reactor—a fast reactor that is cooled by molten lead

light-water reactor—a reactor that uses ordinary water as neutron moderator and as coolant

M

mass number—an integer that is the sum of the number of protons and neutrons in an atom's nucleus

materials test reactor—a reactor typically using aluminum-clad fuel for the primary purpose of performing irradiation tests on materials

microsphere—as used in this report, a very small sphere (about 1 millimeter in diameter) containing a fissile material kernel and several spherical layers of graphite and one of silicon carbide

moderator—a material used in reactors to slow the neutrons produced in fission

molten salt reactor—a type of reactor whose fuel is a molten salt that circulates in a loop in which it is processed

millisievert—one one-thousandth of a sievert

mixer-settler—a device used in separations in which immiscible fluids are mixed and allowed to separate by gravity

N

neutron absorption cross-section—a measure of the probability of a material absorbing a neutron

neutron irradiation—exposure to a source of neutrons

neutron poison—a material that has a high neutron capture cross-section

noble metal—a relatively chemically inert metal, typically having an atomic number of 42 to 46 and 74 to 78

nominal capacity—an assumed or approximately correct capacity

nuclear fuel cycles—the flow of nuclear material from various operations from mining to waste disposal

nuclear fuel recycling center—a site at which more than one of the fuel recycle operations are carried out

O

off-gas—gases and vapors released from equipment, processes, or buildings

P

pellet-cladding interaction—physical contact at the interface between a fuel pellet and its cladding

performance-based—a requirement that is based on meeting a specified goal that does not specify the means of meeting that goal

permselective membrane—a thin layer of a solid substance that is selectively permeable to one or more materials

precipitate (verb)—to form a solid that settles out of a liquid

precipitate (noun)—the material formed during precipitation

precipitation—the process of forming a precipitate (noun)

pressurized water reactor—a reactor that prevents water from boiling at temperatures above its normal boiling point by the application of pressure

probabilistic risk assessment—evaluation of risk incorporating probabilities of something occurring

production reactor—a reactor whose primary function is to produce plutonium or tritium

proliferation—as used in this paper, the undesirable spread of fissile material and/or technology used to produce nuclear weapons

pulse column—a vertical cylinder containing internal structure to disperse and contact two counter-currently or co-currently flowing liquids

pyrolytic graphite—a type of crystalline carbon formed by decomposing organic compounds at high temperatures

pyrolyzed carbon—the amorphous product of decomposition at high temperatures of organic material

pyroprocessing—the act of processing at high temperatures

R

radiation background—the level of radiation found normally in the environment or in a facility containing sources of radiation

radiation dose—the amount of radiation absorbed

radiation shielding—material that diminishes the intensity of radiation

radioelement—an element comprised of radionuclides

radiological hazard—a device or material whose radiation poses a hazard

radionuclide—a radioactive isotope

raffinate—the phase remaining (typically aqueous) after extraction of some specified solute(s) by a solvent (typically organic)

rare earth—synonymous with lanthanide

red oil—a potentially explosive liquid of ill-defined composition formed by the reaction of heat, chemicals, and/or radiation with organic liquids

redox process—an early solvent extraction plutonium separation process employing methyl isobutyl ketone as the extractant

reducing agent—as used in this paper, a chemical capable of chemically reducing another chemical

reenriched—as used in this report, uranium that is isotopically enriched after having been initially enriched, depleted by neutron irradiation, and recovered by reprocessing

refabrication—fuel element fabrication of material recovered in reprocessing

rem—dosage of ionizing radiation that causes the same biological effect as exposure to x rays or gamma rays that produce one electrostatic unit of charge of either sign in 1 cubic centimeter of dry air at 0 °C and 1 atmosphere of pressure; about 100 ergs per gram

remote decontamination—cleaning equipment or facilities without direct contact by operating personnel

remote maintenance—maintaining equipment or facilities without direct contact by operating personnel

repository—as used in this paper, a deep geologic facility for the disposal of wastes

reprocessing—separation of spent nuclear fuel into its constituent components, typically to recover fissile and fertile material

risk-informed—a philosophy whereby risk insights are considered together with other factors to establish requirements that better focus licensee and regulatory attention on design and operational issues commensurate with their importance to public health and safety

robotics—technology dealing with the design, construction, and operation of robots in process automation

S

scrub—process stage in a solvent extraction procedure for removing secondary salt constituents from organic phase before recovery of principal constituent.

self-protecting—an arbitrary classification of protection derived from the radiation dose associated with irradiated spent commercial fuel, generally taken to be the protection afforded by a dose rate of 100 R/h

separation factor—the concentration of the species of interest in the feed to one step of a separation process divided by its concentration in the product of that stage

sievert—the SI unit of absorbed dose equivalent (1 joule/kilogram or 100 rem)

single-shell tank—a radioactive waste tank constituted of only one container

sintering—a process, usually at high temperature, that causes particles of a material to bond into a coherent mass without melting

sludge—a, noncrystalline, mud-like solid material

sol-gel process—a process for producing solids by forming a gel from a colloid

sodium-cooled fast reactor—a fast reactor that is cooled with liquid sodium

solvent extraction—a process wherein a dissolved material is transferred between two contacted immiscible liquids

spent nuclear fuel dissolution—the act of dissolving spent fuel, usually by action of an acid

spent nuclear fuel shearing—the act of cutting fuel elements into pieces to expose the contained fuel material

steam stripping—a process wherein unwanted material is removed from a liquid by passing steam through the liquid

stoichiometry—the numeric relationship of the number of atoms in a chemical compound

stripping—process operation for recovery of constituents extracted into the organic phase in the solvent extraction operation by contacting the organic phase with a dilute acid stream

supernatant liquid—the layer of liquid overlaying a solid such as a sludge

surge capacity—accommodation for storing material awaiting the next step in a process or operation

T

terabecquerel—1+e12 disintegrations per second

thermal neutron spectrum—the range and distribution of neutron energies consistent with the range and distribution of energies of molecules in a gas at room temperature

thorium-uranium fuel cycle—a fuel cycle based on thorium and ^{233}U

transmutation—as used in this report, a process by which one isotope is converted to one or more different isotopes by neutron capture or fission

transuranic actinide isotopes—isotopes whose atomic numbers are greater than 92 and less than 104

tributyl phosphate—an organic compound commonly used in the separation of desired radionuclides, specifically uranium and plutonium, from unwanted radionuclides (e.g., fission products)

TRISO fuel particle—a small fissile fuel particle comprising a fuel kernel and spherical layers of pyrolytically deposited carbon and silicon carbide

U

uncertainty analysis—quantification of the uncertainty in the predication of models

uranium-plutonium fuel cycle—a fuel cycle based on uranium and plutonium

V

vacuum distillation—evaporation of a material at less than atmospheric pressure

valence—a measure of the combining power and ratio of one element or chemical species with another, usually expressed as a small positive or negative integer

very-high-temperature reactor—a reactor that operates at temperatures much above that of light-water reactors

voloxidation—a sequence of oxidation and reduction reactions using some combination of air, oxygen, ozone, hydrogen, and heat to pulverize an oxide fuel

vitrification—production of a glass or glassy substance, commonly used to prepare a high-level waste form

W

water scrubber—a device that uses water to remove impurities from a gas by intimate contact with the water

Z

zeolite—a crystalline silicate with internal cavities large enough to accommodate atoms and small molecules; commonly used in separations, especially of gases

Zircaloy—an alloy, primarily composed of zirconium alloyed with minor elements such as tin, used in the form of tubes (cladding) to contain fuel pellet

THIS PAGE IS INTENTIONALLY LEFT BLANK.

APPENDIX A: DESCRIPTION OF THE BARNWELL NUCLEAR FUEL PLANT DESIGN AND PUREX PROCESS

This appendix describes in some detail the reprocessing of spent nuclear fuel, including the PUREX process, based on the last attempt to build and operate a reprocessing plant (the Barnwell Nuclear Fuel Plant (BNFP)) in the United States. Many changes and improvements have been made since the mid-1970s when the BNFP was under construction. The following description illustrates the state of the art in reprocessing at that time. In general, the principal process steps are the same today as they were then.

Spent Fuel Receiving and Storage

The irradiated fuel assemblies would arrive at the reprocessing plant on a carrier in shielded casks. The cask and carrier would be monitored for external contamination and washed to remove external dirt. After the cask had been removed from the carrier, the condition of the fuel and cask would be determined. The cask would be vented, cooled, and prepared for entry into the cask unloading pool. The cooled cask would be moved by the cask-handling crane to the cask unloading pool, where it would be lowered to the bottom of the pool. The top of the cask would be opened, and the contained fuel would be removed. The identity of each fuel assembly would be established and compared against shipping documentation. The fuel would be placed in storage canisters, which would be moved to the fuel storage pool for retention until the fuel was scheduled for reprocessing. All operations would be performed under water.

Spent Fuel Inventory

A typical 1500 MT per year of uranium per year (MTU/yr) reprocessing/recycling facility will generally have a spent fuel storage capacity of approximately 2000 fuel elements, which, depending on the burnup, will represent approximately one-fourth of the annual plant capacity (e.g., the BNFP could store 360 MTU at any one time). Table A1 shows initial BNFP specifications for spent fuel in the mid-1970s.

Table A1: Spent Nuclear Fuel Specifications circa the Mid-1970s

Characteristic	Value
Burnup, maximum	40,000 MWd/MTU
Specific power, maximum	50 MW/MTU
Enrichment	Initial: 3.5–5.0% U-235 or equivalent Final: 1.9–3.5% U-235 + Pu content
Plutonium yield, total	10 kg Pu/MTU
Age of spent fuel, as shipped	90-day cooled, minimum
Age of spent fuel at start of reprocessing	90-day cooled, minimum
Cladding	Zircaloy or stainless steel
Maximum dimensions	11-3/8" sq. by 20' long

At the current time, however, initial feed spent fuel will be aged for years (some for as long as 40 years) since the electric utilities continue to store the fuel.

Shearing and Dissolving

An individual spent fuel assembly container would be remotely transferred from the storage pool, and the individual fuel assemblies would be removed and moved to the feed mechanism of the mechanical shear. Generally, a full batch or a lot of fuel from a single source would be processed at one time. The fuel assemblies would be chopped into small segments (approximately 2 to 5 inches long) to expose the fuel to the nitric acid dissolver solution.

The chopped fuel assemblies would fall into one of three dissolvers that contain hot 3–molar (M) HNO_3 to dissolve virtually all uranium, plutonium, other actinides, and most of the fission products. During dissolution, a soluble neutron poison (gadolinium nitrate) would be added to the dissolver as a precaution to prevent a criticality. After the initial dissolution, a digestion cycle would be used (8-M HNO_3) to dissolve any remaining fuel (MOX fuel is sometimes refractory and requires more aggressive dissolution conditions). Following digestion in nitric acid, any remaining insoluble material would be rinsed with dilute nitric acid, and these materials plus the undissolved cladding hulls of stainless steel or Zircaloy would remain in the dissolver basket. Gases released from the spent fuel during dissolution (primarily ^{85}Kr, tritium, ^{129}I, and $^{14}CO_2$, with the possibility of some $^{106}RuO_4$) and nitrogen oxides would be directed to the off-gas treatment system to remove particulates, radioiodine, and nitrogen oxides. The cladding hulls would be rinsed, monitored for fissile material, packaged, and transferred to the solid waste storage area. The nitrogen oxides would be reconstituted to nitric acid.

Product Separation and Purification

After acidity and concentration adjustment, the dissolver solution would become the solvent extraction process feed solution. It would be clarified by centrifugation and then sent to the first solvent extraction decontamination cycle. In this cycle, the feed solution is contacted counter-currently in a 10-stage centrifugal contactor with an organic solution of 30-percent tributyl phosphate (TBP) in a kerosene or normal paraffin hydrocarbon diluent (primarily dodecane). The organic solution preferentially would extract the tetravalent plutonium and hexavalent uranium, leaving about 99 percent of the fission products in the aqueous raffinate (waste) nitric acid solution. The organic solution from the centrifugal contactor then would pass through a pulsed scrub column where aqueous 3-M HNO_3 solution scrubs (back-extracts) about 96 percent of the small amount of extracted fission products from the product-bearing organic solution. This scrub solution subsequently would be recycled to the centrifugal contactor for additional uranium and plutonium recovery to reduce the potential for product losses. The combined aqueous stream leaving the centrifugal contactor would contain approximately 99.6 percent (or more) of the fission products and would be sent to a high-level waste (HLW) concentrator.

The organic solution from the scrub column (joined by organic raffinates from downstream plutonium purification columns) would pass through a partitioning column where tetravalent plutonium would be electrochemically reduced[30] to the less extractable trivalent state. This would enable the plutonium to be stripped quantitatively into an aqueous nitric acid solution within the electrochemical unit. A substantial amount of uranium would follow the plutonium in the aqueous stream (some uranium is also electrolytically reduced from U(VI) to U(IV) and may in fact be the ultimate plutonium reductant). The aqueous stream, which is approximately 35-percent plutonium and 65-percent uranium, would flow to the plutonium purification cycles. The organic solution, now stripped of plutonium, would pass through another pulsed column where the residual uranium would be stripped into a weakly acidified aqueous solution (approximately 0.01-M HNO_3).

The aqueous strip solution containing the residual uranium would be concentrated by evaporation from 0.3-M uranium to 1.5-M uranium and adjusted with nitric acid to approximately 2.5-M HNO_3. This uranium would be preferentially extracted again by a 30-percent TBP organic solution in another pulsed column. Before leaving the column, the organic solution would be scrubbed with dilute nitric acid solution, which would remove traces of extracted ruthenium and zirconium-niobium fission products, which are among the fission products most difficult to remove. Hydroxylamine nitrate or hydrazine also would be added to the scrub solution to remove residual plutonium by its chemical reduction to the inextractable trivalent state. Uranium subsequently would be stripped from the organic solution in another pulsed column, using an acidified aqueous solution (0.01-M HNO_3). This solution would be concentrated, by evaporation, from 0.4-M uranium to 1.5-M uranium. Finally, the concentrated aqueous uranium solution would be passed through silica gel beds to remove residual traces of zirconium-niobium fission products, and the uranyl nitrate product solution would be analyzed and transferred to the UF_6 facility for storage or conversion to UF_6 and subsequent shipment. Uranium recovery was expected to be at least 99 percent. Removal of fission products was to be 99.99 percent.[31]

Plutonium in the aqueous stream leaving the partitioning column would be re-oxidized to the organic-extractable tetravalent state by sparging the solution with di-nitrogen tetroxide (N_2O_4) and would be preferentially extracted into an organic solution in the first pulsed extraction column of the second plutonium cycle. In the top portion of this column, the organic stream would be scrubbed with 10-M HNO_3 solution to remove traces of extracted ruthenium and zirconium-niobium fission products. The organic stream then would pass through a strip column where tetravalent plutonium would be transferred to an aqueous stream of dilute (0.3-M) nitric acid. This cycle would also partition plutonium from the accompanying uranium, with the uranium being recycled. The extraction-scrubbing sequence would be repeated in a third plutonium cycle for further decontamination from fission products and uranium. To effect a higher plutonium product concentration, the plutonium would be reduced in the third-cycle strip column by hydroxylamine nitrate to the more hydrophilic trivalent state. A TBP organic scrub solution would be added to remove any residual uranium from the plutonium aqueous stream as it leaves the third-cycle strip column. Following the third plutonium cycle, the plutonium nitrate solution would be washed with a stream of organic diluent in a final column to remove traces of organic solvent (TBP). Final plutonium concentration would be established in a critically-safe-geometry

---

[30] Electrochemical reduction of plutonium was unique to the Barnwell plant. Plutonium is conventionally reduced chemically, often with U(IV).

[31] For a description of actual operating experience at the THORP, see Section 3.1.3.

evaporator made of titanium. The plutonium product solution would be analyzed and stored in critically safe tanks. The plutonium recovery was expected to be 98.75 percent.

The contaminated organic solvent stream from the co-decontamination and partition cycles would be washed successively with dilute aqueous solutions of sodium carbonate, nitric acid, and sodium carbonate to remove organic degradation products (primarily dibutyl and monobutyl phosphate) generated by radiation damage to TBP. This step would produce waste solids formed from the sodium salts and organic phosphates.

The precipitated solids would be removed by filtration following the first carbonate wash. Fresh TBP and/or diluent would be added, as required, to maintain the 30-percent TBP concentration and the total solvent inventory at the desired level. The contaminated organic solvent stream from the second uranium cycle would be treated similarly in a separate system, except that the second sodium carbonate wash would be omitted.

The aqueous raffinate streams from the plutonium and uranium cycles, except for the last product-bearing raffinate, would be treated with N_2O_4 for adjustment of the plutonium oxidation state to Pu(IV) and U(VI) and would be passed through a pulse column where residual uranium and plutonium would be recovered by extraction into a 30-percent TBP organic solution. The recovered uranium and plutonium would be recycled back to the decontamination cycle for recovery. The aqueous raffinate stream would be concentrated in a low-activity process waste evaporator.

Liquid Waste Streams

The radioactive aqueous waste streams from all the solvent extraction cycles would be concentrated in the high- or low-activity waste evaporators, depending on the relative radioactivity content. The acidic concentrated high-level liquid waste bottoms would be stored in a cooled stainless steel waste tank. The evaporator overheads would be passed through a distillation column to recover the nitric acid as a 12-$\underline{M}$ solution. The distillation column overhead (primarily water) then would be recycled as process water, or sampled and released to the stack from a vaporizer provided it met release specifications. The recovered 12-$\underline{M}$ HNO_3 would be used in parts of the process where the residual radioactivity could be tolerated.

Miscellaneous aqueous streams containing salts and fission products (approximately 1 curie per liter (1 Ci/L) but no appreciable uranium or plutonium would be acidified and concentrated to approximately 50 Ci/L in the general purpose evaporator. These evaporator bottoms would be stored in an uncooled stainless steel waste tank. The condensed overheads would be vaporized to the stack.

Process Off-gas Streams

Off-gases from the dissolver would be scrubbed with a mercuric nitrate solution to reduce levels of radioactive iodine in the effluent and then treated in an absorber to convert nitrogen oxides to nitric acid suitable for recycling. The dissolver off-gas and vessel off-gas streams would be combined and passed successively through a second iodine scrubber containing mercuric nitrate, silver zeolite beds for iodine sorption, and high-efficiency filters before release to the stack.
Facilities for the retention of other radionuclides such as [85]Kr, tritium, and [14]C (as CO_2) were not in place in the 1970s reprocessing plant, although there were plans to recover [85]Kr.

UF$_6$ Preparation

The UF$_6$ plant was designed with an annual capacity of 1500 MTU and assumed to operate 24 hours per day for up to 300 days a year. Scrap from the plant operations would be stored until processed in the appropriate facility after which it would be shipped off site for either re-use or for disposal as contaminated waste, as determined by analysis.

The individual process steps for the conversion of uranyl nitrate to uranium hexafluoride in a UF$_6$ conversion plant co-located with a reprocessing/recycling facility are the following:

- receipt of purified uranyl nitrate solution from a reprocessing plant
- concentration of the uranyl nitrate feed solution via evaporation
- conversion of the uranyl nitrate to UO$_3$ by heating to denitrate it
- hydrogen reduction of UO$_3$ to UO$_2$
- hydrofluorination of UO$_2$ to UF$_4$, using gaseous HF
- fluorination of UF$_4$ to UF$_6$, using electrolytically generated F$_2$
- freezing and then resubliming UF$_6$ in a series of cold traps to purify it[32]
- packaging of the UF$_6$ product into standard transport cylinders

All processing steps that involve radioactive materials would be performed inside equipment maintained at negative pressure relative to the adjacent, less radioactive areas of the conversion building. The pressure differences would be maintained so that air flow is from uncontaminated areas into areas of potentially higher contamination levels, thus limiting the spread of radioactivity.

The equipment would form the first level of confinement; the conversion building would form the second level. Pressure differences would be maintained by automatically controlled, zoned ventilation systems. Spare ventilation fans and required controls, which are provided, would be connected to independent or installed emergency power systems in the event of loss of normal plant power, to ensure maintenance of the required pressure differences.

Plutonium Precipitation and Conversion

The feed material for the plutonium product facility (PPF) would be separated plutonium nitrate solution from the plutonium nitrate storage tanks in the separations facility. Table A2 gives its typical characteristics. The alpha, neutron, and gamma emissions would require special features for confinement and shielding. The radioactive decay heat and potential criticality of concentrated plutonium solutions and products would require special design constraints for the processing equipment within the PPF.

---

[32] Small amounts of some radionuclides having volatile fluorides, most notably tellurium, neptunium, and technetium, follow the uranium all the way to the UF$_6$ plant and must be removed by fractional sublimation.

Table A2: Expected characteristics of Plutonium Nitrate Feed to the BNFP Plutonium Product Facility

Characteristic	Value
Plutonium concentration, g/L*	100–360
Nitric acid concentration, M	2–10
Uranium concentration, ppm	Less than 10,000
Radioactive decay heat, Btu hr^{-1} (kg Pu)$^{-1}$	Less than 60
Radioactive hydrogen generation, scfh/kg Pu	Less than 5x10^{-4}
Gamma emissions, Ci/g Pu	80
Pu-238, % of total Pu	2.5
Pu-239, % of total Pu	50
Pu-240, % of total Pu	25
Pu-241, % of total Pu	15
Pu-242, % of total Pu	7.5

*Plutonium concentrations in excess of 250 g/L may be processed if the heat generation rate is less than 60 Btu hr^{-1} (kg Pu)$^{-1}$.

The plutonium nitrate solution would be transferred from the storage tanks to one of two feed preparation tanks on a batch basis. The nitric acid concentration would be adjusted to 3.0 M to provide a constant feed for the conversion process. The concentration must be maintained at more than 2 M to ensure the prevention of plutonium hydrolysis which can form plutonium colloid (polymer formation) and oxide precipitation. Hydroxylamine nitrate (HAN) also would be added at the feed adjustment tank to reduce any Pu(VI) to Pu(IV) before the precipitation step.[33] After completion of the feed adjustment step, the plutonium nitrate solution would be heated to 60 °C in an inline heater and fed continuously into a precipitator equipped with a mechanical stirrer. A solution of 1.0-M oxalic acid would be added to the precipitator, and the resulting plutonium oxalate slurry would be allowed to overflow to the digester, the role of which is to grow large, well-formed crystals. The digester would consist of three inline mechanically stirred vessels (identical to the precipitator) that would be arranged to permit the overflow of one unit to cascade into the next. The precipitation and digestion vessels would be sized such that the residence time is approximately 1 hour.

The slurry would be fed into a rotary-drum vacuum filter for liquid-solid separation. The oxalate cake would be rinsed on the filter drum and scraped off with a "doctor blade." The filtrate would be transferred to a filtrate surge tank before further processing. The plutonium oxalate cake from the drum filter would be discharged directly into a rotary screw dryer-calciner. The oxalate anion would be destroyed by heating in air to form the desired plutonium dioxide product. The oxalate-cake feed rate, residence time, heating rate, and final calcining temperature are all critical to the production of a plutonium dioxide feed material with the proper characteristics for manufacturing into satisfactory fuel pellets during subsequent mixed-oxide (MOX) fuel fabrication operations. [*The reader should recognize that this process was optimized for the anticipated MOX fuel specifications of the mid-1970s. The final product specifications required for a fuel in 2010 or later will establish the ultimate plutonium conversion process.*]

---

[33] The electrochemical potentials of the various plutonium valence states are such that Pu(III), Pu(IV), and Pu(VI) can coexist in solution at equilibrium. Consequently, it is necessary to chemically produce the desired valence state.

The calciner would discharge directly onto a continuously moving screen. The powder passing through the screen would be collected in a geometrically safe blender body, which has a maximum capacity of 40 kilograms of plutonium as plutonium oxide. The oversize product would pass off the top of the screen into a collection hopper. This hopper would be periodically emptied into a grinder which would reduce the particle size to meet the product specification. The grinder would empty into an identical 40-kilogram blender. The ground plutonium oxide would be recycled to either the top of the screen or to the dryer-calciner. These operations are especially "dirty" in that they produce a plutonium dioxide dust that is difficult to contain and handle.

Plutonium Sampling and Storage

A blender would receive nominally 32 kilograms of plutonium oxide, as indicated by a weighing element beneath the blender. To change the vessel, it would be remotely valved off and transferred to the blending stand. The full blender body would be rotated about its radial center until completion of blending. The powder would be sampled and the samples analyzed to determine properties and insure homogeneity. The plutonium would be held in the blender body until the analytical results were received. Plutonium dioxide not meeting the product specifications would be either recycled or loaded out and held for future rework.

The blended powder in the blender body would be transferred to the powder loadout stand where the contents of the blender would be discharged into four product canisters, each holding nominally 8 kilograms of plutonium oxide. The canister covers would be installed, each canister would be sealed, and the outer surface would be decontaminated. Four product canisters would be loaded into a pressure vessel that would double as a storage container and primary containment vessel during shipment. The pressure vessels (which were never built) were to be vented through a three-stage high-efficiency particulate air filter. The loaded pressure vessel would be placed either in the storage vault or into a shipping container for offsite shipment (if the MOX fuel fabrication plant were colocated with the reprocessing/recycling facility, offsite shipment would not be necessary).

Recycle Streams

Filtrate from the vacuum drum would be collected in the filtrate surge tank where gas and liquid would be separated. The gas would be routed to the vacuum pump. The majority of the gaseous output of the vacuum pump would be recycled to the vacuum drum filter. A small amount of the gas would be bled to the vessel off-gas system.

The liquid from the filtrate surge tanks would be pumped through cartridge-type secondary filters into the filtrate evaporator feed tank. The filtrate would be transferred from the filtrate evaporator feed tank by air lift into the filtrate evaporator. In the evaporator, the filtrate would be distilled sufficiently for destruction of the oxalic acid and to reduce the volume of solution containing plutonium.

The residue from the evaporator would be sequentially cooled, passed through another secondary cartridge-type filtration step to remove any possible solid (normally not expected), and then collected in the concentrate catch tank. The filtrate concentrate then would be transferred by jet to the concentrate sample tank where it would be sampled. If analyses indicated the presence of oxalic acid, it could be destroyed by returning the concentrate to the filtrate evaporator feed tank for reprocessing or by adding acidified potassium permanganate to the sample tank. The contents of the sample tank also would be returned to the evaporator feed tank if the presence of solids containing plutonium was detected. When sampling indicates the plutonium content/mixture is satisfactory, the concentrate would be transferred to a storage tank from which it would be pumped to the separations facility for plutonium recovery.

The evaporator overhead would be condensed, combined with condensate from the off-gas system, and filtered with cartridge-type filters to remove any possible solids. The distillate would be collected in the distillate catch tank from which it would be transferred batchwise to the distillate sample tank. Depending on analyses, the distillate could be transferred to the evaporator feed tank for reprocessing, the concentrate storage tank when containing recoverable plutonium, or the distillate storage tank. From the distillate storage tank, the distillate could be transferred to the separations facility for acid recovery.

Waste Treatment

A typical commercial reprocessing/recycling plant of the 1970s generated gaseous, liquid, and solid waste, as would any modern day plant. Continuing with the example of the 1500 MTU/yr designed separations capacity of the BNFP, the waste treatment specifications were as follows.

Low-Level Liquid Wastes

At the BNFP, low-level aqueous liquid waste was planned to be released into local area streams at the rate of about 2000 gallons per minute (at full nominal rated operation). Maximum release temperature was 85 °F with essentially no radioactivity and only water treatment chemicals in the water.

High-Level Liquid Wastes

High-level liquid waste was to be solidified after a minimum of 5 years of tank storage and transported to a Federal repository within 10 years of generation. The BNFP initially constructed two 300,000-gallon storage tanks, manufactured of 304L stainless steel, double-walled and designed with internal stainless steel cooling coils. Relevant design data on tank contents are noted below:

- activity: 1.80×10^4 Ci/gal
- acid concentration: 1–5 M HNO_3
- temperature: 140 °F
- heat generation rate: 72,000 BTU/h·MTU

Each cylindrical high-level liquid waste tank was 16.5 meters in diameter by 6.1 meters high and was contained within an underground cylindrical concrete vault lined with stainless steel. Each vault was 18.3 meters in diameter and 7.6 meters high. The vault floor, walls, and top were 1.2 meters, 0.9 meters, and 1.7 meters thick, respectively.

It was anticipated that three additional 300,000-gallon tanks would need to be constructed for a total capacity of 1.5 million gallons. This was expected to allow for ample storage of liquid waste before solidification and offsite shipment to the Federal repository (not identified at that time). Each high-level liquid waste tank contained the following equipment:

- 48 5-centimeter-diameter cooling coils
- 18 air-operated ballast tanks around the perimeter of the tank
- 9 air-operated ballast tanks in the main part of the tanks
- 22 air-lift circulators
- 5 steam-operated ejector pumps (empty-out jets)
- water-seal type pressure/vacuum relief system
- multiple external temperature sensing points
- 10 instrument dip tubes to measure liquid level and specific activity

The waste solidification plant (WSP) would contain the waste vitrification equipment, canister sealing, inspection and decontamination equipment, off-gas treatment equipment, and remote maintenance facilities in four process cells. Table A3 presents the primary process functions that would be performed in each of the cells. All process cells in the WSP would be completely lined with stainless steel. The cells were to be surrounded by limited access areas for operating and controlling the processes in the cells. All operational and maintenance functions in the process cells would be performed remotely using viewing windows, manipulators, and cranes.

Table A3: Function of Cells in the BNFP Waste Solidification Plant

Cell Description	Cell Function
Waste vitrification	Calcine liquid waste; vitrify calcined waste; weld canisters closed
Canister decontamination	Remove external radioactivity from the canister
Off-gas treatment	Treat off-gas from WSP process vessels
Hot maintenance	Perform remote maintenance on contaminated equipment

Solid Waste Disposal

Solidified HLW, hulls, and alpha wastes were to be stored on site in an interim storage area with eventual transport to a Federal HLW repository. Spent fuel hull treatment was to be optimized (e.g., hulls would be compacted or melted) to minimize overall capital and/or operating costs. Because of the BNFP site location, transport may have been by truck or rail or by intermodal means (including barge from site to port and thence by rail or truck to the repository).

Low-level solid waste would be disposed of at a licensed low-level waste facility. At the BNFP facility, such disposal was simplified as the Chem-Nuclear Barnwell low-level waste site was immediately adjacent to the facility. While minimizing transportation costs, the facility would have had to meet all other relevant regulatory requirements.

Off-Gas System

For the principal plant off-gases, the initially projected release rates were the following:

- Iodine
  ^{129}I: 1.4×10^{-6} Ci/s (99.9% + % retained in plant)
  ^{131}I: 1.1×10^{-5} Ci/s (99.9% + % retained in plant)

- Krypton
  ^{85}Kr: 4.3×10^{-1} Ci/s (no recovery facilities were planned in the design being initially licensed)

- Tritium
  ^{3}H: 1.8×10^{-2} Ci/s (no recovery facilities were planned in the design being initially licensed)

- NOx
  200 lb/h (release concentration less than 150 ppm (at top of stack))

At the time, these releases were acceptable. However, as these earlier designs proceeded through their review, agreement with the license was reached with the Council on Environmental Quality that an effort would be made to minimize krypton and tritium releases, even though capturing these gases was not required then.

Cryogenic systems were considered and were being evaluated until the International Nuclear Fuel Cycle Evaluation started and the concomitant ban on reprocessing was invoked, which halted further commercial reprocessing development activity.

Nuclear Material and Quality Control Groups

The facility organization will normally include a nuclear material control group which will have responsibility for developing and carrying out an accounting plan. In addition, a facility is likely to have an independent quality control group to assure compliance with the requirements imposed on the facility.

The basic accounting method developed at Nuclear Fuel Services (NFS) is conventional material balance accounting. The facility is divided into a number of material balance areas, and all of the movements of materials into and out of these areas are measured and recorded. At periodic intervals, the inventory of materials in each of the areas is measured, and a material balance is "closed." During each material balance period, the sum of the initial inventory in an area and the inputs during the period should equal the sum of the final inventory and outputs. Any discrepancy is labeled as "material unaccounted for" or "inventory difference." If the discrepancy exceeds values that might be expected to result from measurement uncertainties, then further measures are undertaken to attempt to identify the source or sources.

The material balance areas used for internal accounting purposes may not coincide with those required for national or international safeguards systems. At NFS, the following eight areas were designated for internal accounting:

(1)   Fuel Receiving and Storage
(2)   Mechanical Processing and Dissolution
(3)   Input Accountability and Feed Adjustment
(4)   Process Product Storage
(5)   Shipment
(6)   Waste Treatment
(7)   Underground Waste Storage
(8)   Analytical Laboratories

Several of these areas may be treated as one for accounting purposes under national or international safeguards systems.

Records

- Fuel Receipt Form—This form includes information on each fuel assembly provided by the shipper including calculated uranium and plutonium content based on fuel fabricator and reactor operating history data.

- Fuel Storage Record—This record is the canister number and pool storage location of each fuel assembly received. This information is also maintained on a status board.

- Feed Magazine Loading Record and Shear Operating Record—These record the movement of material within the process mechanical cell and removal of assemblies from storage.

- General Purpose Cell Record—This is used to record the storage of chopped fuel, movement to dissolution, and any pumping from the cell sump.

- Leached Hull Record—This records the gross weight, tare weight, net weight, sampling code, and removal date of drums containing leached hulls.

- Input Accountability Record—This form records the instrument readings for the input accountability tank loading and the input sample identification.

- Liquid Waste and Product Storage Tank Measurements—These record the instrument readings and sample identification for the various process accountability vessels.

- Plutonium or High-Enrichment Uranium Product Load-out Record—This records the gross and net weights of the product load-out containers, as well as their storage locations.

- Analytical Services Form—This records the sample identification and analytical results.

- Inventory Record—This form is used to record the instrument readings and sample identification for in-process material in various vessels at the end of each processing campaign.

- Material Status Report—This is a consolidated inventory record which is prepared every 3 months.

- Shipping Form—This is used to record accounting data on material shipped from the NFS.

Analyses on Accountability Samples

The analyses performed on accountability samples include total uranium, total plutonium, isotopic plutonium, isotopic uranium, and density. The techniques used include mass spectrometry, amperometric titration, isotopic dilution, alpha counting, high-resolution gamma spectroscopy, and various other chemical analysis techniques. The analysis of samples from the input accountability and feed adjustment tank are particularly important for accounting purposes.

The input plutonium concentration is determined by an isotopic dilution technique. The input samples are diluted, spiked with ^{242}Pu (or sometimes ^{244}Pu), purified by ion exchange, and then analyzed with a mass spectrometer. To determine the isotopic weight percentages, unspiked samples are analyzed with a mass spectrometer. Similar techniques are used for uranium measurements, but ^{235}U or ^{233}U is used as the spiking isotope. Similar techniques are used for the assessment of output solutions with the exception that titration techniques are normally used to determine the uranium and plutonium concentrations. In addition to assessment of the product solutions, the waste materials are also assayed for uranium and plutonium content.

THIS PAGE IS INTENTIONALLY LEFT BLANK.

APPENDIX B
DECAY HEAT IN SPENT FUEL

Figure B1 [Croff, 1982)] shows the contributions of selected actinides and fission products to heat generation rate from SNF in waste as a function of decay time for fuel irradiated to 33 gigawatt-day per metric ton initial heavy metal (GWd/MTIHM). It is noteworthy that the decay heat from the actinides in SNF (^{241}Am and $^{238, 240}$Pu) exceeds that of the fission products (primarily ^{90}Sr and ^{137}Cs and their progeny) after a decay time of about 60-70 years [Roddy, 1986].

Reprocessing relatively short-cooled spent fuel has advantages and disadvantages. Advantages accrue with respect to decay heat reduction in the wastes because actinides and selected fission products are removed before storage and disposal of the wastes. Advantages relate to reducing the volume of spent fuel stored which reduces the need for spent fuel storage facilities and storage casks and the repository volume required for the HLW resulting from reprocessing the SNF, and reduces the potential risk of proliferation or from terrorist attack on the stored spent fuel. The relative reduction in long-term decay heat production decreases as the SNF gets older before reprocessing because more ^{241}Pu decays to ^{241}Am.

The disadvantages of reprocessing relatively short-cooled spent fuel are related to the necessity of handling more highly radioactive fuel, which increases the potential hazards and add to the complexity and cost of the reprocessing plant and processes.

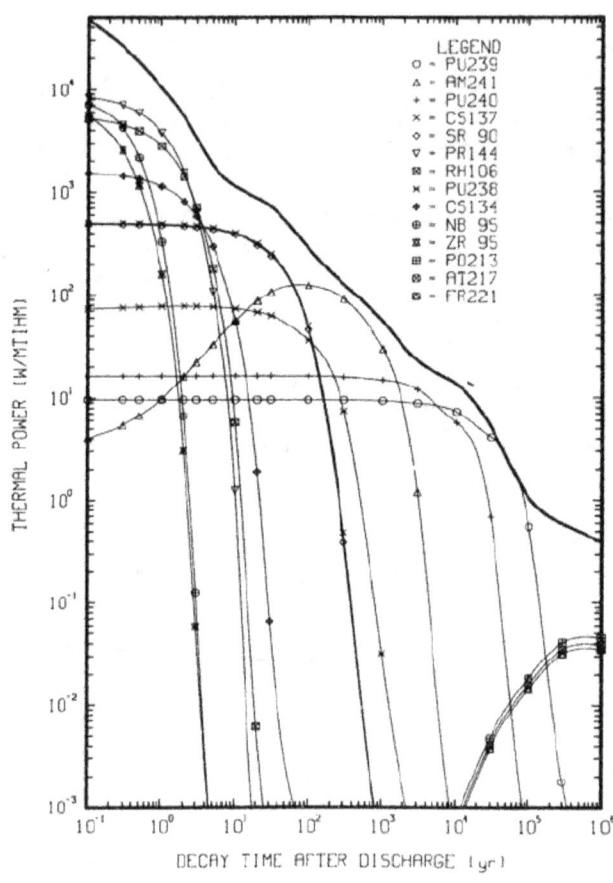

Figure B1: Contributions of selected actinides and fission products to SNF decay heat generation rate

THIS PAGE IS INTENTIONALLY LEFT BLANK.

APPENDIX C
COMMITTEE LETTERS RELATED TO RISK-INFORMED ACTIVITIES AND PROBABILISTIC RISK ASSESSMENT

- ACNW&M letter dated May 2, 2006, from Michael T. Ryan, Chairman, ACNW&M, to Nils J. Diaz, Chairman, NRC, Subject: Risk-Informed Decision-Making for Nuclear Materials and Wastes.

- ACNW&M letter dated May 3, 2004, from B. John Garrick, Chairman, ACNW&M, to Nils J. Diaz, Chairman, NRC, Subject: Risk Insights Baseline Report.

- ACNW&M letter dated August 13, 2003, from B. John Garrick, Chairman, ACNW&M, to Nils J. Diaz, Chairman, NRC, Subject: High Level Waste: Risk-Significance Ranking of Agreements and the Use of Risk Information to Resolve Issues.

- ACNW&M letter dated July 2, 2002, from George M. Hornberger, Chairman, ACNW&M, to Richard A. Meserve, Chairman, NRC, Subject: The High-Level Program Risk Insights Initiative.

- ACNW&M letter dated April 29, 2002, from George M. Hornberger, Chairman, ACNW&M, to William D. Travers, Executive Director for Operations, NRC, Subject: Response to Letter Dated March 6, 2002, Concerning Risk-Informed Activities in the Office of Nuclear Material Safety and Safeguards.

- ACNW&M letter dated January 14, 2002, from George M. Hornberger, Chairman, ACNW&M, to Richard A. Meserve, Chairman, NRC, Subject: Risk-Informed Activities in the Office of Nuclear Material Safety and Safeguards.

- ACNW&M letter dated June 29, 2001, from B. John Garrick, Chairman, ACNW&M, to Richard A. Meserve, Chairman, NRC, Subject: Risk-Informed, Performance-Based Regulation of Waste Management and Decommissioning.

- ACNW&M letter dated July 27, 2000, from B. John Garrick, Chairman, ACNW&M, to Richard A. Meserve, Chairman, NRC, Subject: Development of Risk-Informed Regulation in the Office of Nuclear Material Safety and Safeguards.

- ACNW&M letter dated March 26, 1998, from B. John Garrick, Chairman, ACNW&M, to Shirley Ann Jackson, Chairman, NRC, Subject: Risk-Informed, Performance-Based Regulation in Nuclear Waste Management.

- ACNW&M letter dated October 31, 1997, from B. John Garrick, Chairman, ACNW&M, to Shirley Ann Jackson, Chairman, NRC, Subject: Application of Probabilistic Risk Assessment Methods to Performance Assessment in the NRC High-Level Waste Program.

THIS PAGE IS INTENTIONALLY LEFT BLANK.

APPENDIX D
10 CFR PART 55, "OPERATORS' LICENSES"

As written, Title 10, Part 55, "Operators' Licenses," of the *Code of Federal Regulations* (10 CFR Part 55) applies to utilization facilities (e.g., nuclear reactors) and not to reprocessing plants. Key provisions in the regulation concerning operator's licenses are as follows:

"(a) The applicant shall:

(1) Complete NRC form 398, "Personal Qualification Statement – Licensee," which can be obtained by writing the Office of Information Services, U.S. Nuclear Regulatory Commission, Washington, D.C. 20555-0001, by calling (301) 415-5877, or by visiting the NRC's Web site at http:www.nrc.gov and selecting forms from the index found on the home page;

(2) File an original of NRC Form 398, together with the information required in paragraphs (a) (3), (4), (5) and (6) of this section, with the appropriate Regional Administrator;

(3) Submit a written request from an authorized representative of the facility licensee by which the applicant will be employed that the written examination and operating test be administered to the applicant;

(4) Provide evidence that the applicant has successfully completed the facility licensee's requirements to be licensed as an operator or senior operator and of the facility licensee's need for an operator or a senior operator to perform assigned duties. An authorized representative of the facility licensee shall certify this evidence on Form NRC-398. This certification must include details of the applicant's qualifications, and details on courses of instruction administered by the facility licensee, and describe the nature of the training received at the facility, and the startup and shutdown experience received. In lieu of these details, the Commission may accept certification that the applicant has successfully completed a Commission approved training program that is based on a systems approach to training and that uses a simulation facility acceptable to the Commission under Section 55.45(b) of this part;

(5) Provide evidence that the applicant, as a trainee, has successfully manipulated the controls of either the facility for which a license is sought or a plant-referenced simulator that meets the requirements of 55.46c. At a minimum, five significant control manipulations must be performed that affect reactivity or power level (this requirement is obviously directed to operating a nuclear reactor, not to a reprocessing plant). Control manipulations performed on the plant-referenced simulator may be chosen from a representative sampling of the control manipulations and plant evolutions described in 55.59 of this part, as applicable to the design of the plant for which the license application is submitted. For licensed operators

D-1

applying for a senior operator license, certification that the operator has successfully operated the controls of the facility as a licensed operator shall be accepted; and

(6)   Provide certification by the facility licensee of medical condition and general health on Form NRC-396, to comply with Sections 55.21, 55.23 and 55.3(a)(1)."

A copy of NRC Form-398 is to be attached for information.

These requirements have evolved over the past several decades and are much more detailed than the 10 CFR Part 55 rules in existence 30 years ago. It should also be noted that at this time there is no "simulation facility acceptable to the Commission" for a commercial reprocessing/ recycling plant.

THIS PAGE IS INTENTIONALLY LEFT BLANK.

APPENDIX E
RADIONUCLIDE DISTRIBUTION AMONG UREX+1A PROCESS STREAMS

The following discussion presents the assumptions used to specify the paths followed by the elements in an ORIGEN2 [Croff, 1980] calculation that models the UREX+1a process streams. References to the literature sources used are presented where they are available and notes on final output stream characteristics assumptions are presented.

The basis for the following discussion is 1 metric ton initial heavy metal (MTIHM) of pressurized water reactor (PWR) fuel irradiated to 33 gigawatt-days (Gwd)/MTIHM and cooled 25 years. The composition of the initial SNF is described in [Croff, 1978]. The results of the ORIGEN2 calculation are documented in the Agencywide Documents Access and Management System (ADAMS) Accession No. ML072820458.

I. Dissolver Solids, Tc, Volatile Elements, and Cladding

A. Composition of Solids in Dissolver after Dissolution [Campbell, 2007; Kleykamp, 1984; Forsberg, 1985]

Element	Range, % of element in SNF matrix
Tc	8 -12 (assume 15 %)
Ru	27-47 (assume 50 %)
Pd	10-18 (assume 20 %)
Mo	16-41 (assume 40%)
Rh	6-11 (assume 10 %)

Note: percentages tend to increase with burnup

B. Distribution of Technetium Among Process Streams

15% of the Tc is in the dissolver solids and 85% in the dissolver solution (see above)

The concentration of ^{99}Tc in the final U product is based on typical measured values at THORP (see Table 3 in the main report) which is 0.03 ppmw. This equates to a fraction of 0.0198 of the Tc in the dissolver solution being in the uranium product assuming 50% of the Tc in the first cycle solvent extraction product is removed by uranium cleanup

By difference, 99.7794% of the Tc in the dissolver solution is in the Tc product stream that goes to Tc recovery and then the Tc waste form.

The Tc proceeding through the process after the first cycle is assumed to split 50:50 between the TRU product and the fission product waste.

C. Volatiles and Gases [Mineo, 2002]

Use of voloxidation was assumed

^{3}H

Tritium present after a 25y decay is assumed to be recovered by voloxidation [Goode, 1973b] in a closed system with zero external water present. 100% recovery assumed. Basis:
- ORNL/TM-3723 [Goode, 1973a] reported less than 0.1% of T remained in fuel matrix after voloxidation
- In theory, T in the form of ZrT_2 should be dissociated because this occurs at ~300 C [OSHA, 2007] whereas voloxidation occurs at 450 C or higher and hardware melting occurs at 1450 C so the T should be evolved. Experimental information on this is non-existent.

Fraction assumed to be captured in off-gas system: 1.00

Kr

Fraction assumed to be released during voloxidation and dissolution: 1.00

Fraction of Kr released to off-gas system that is captured: 0.85 [EPA, 1977].

Iodine

Fraction to off-gas system from voloxidation: 0.01 [Vondra, 1977a,b]

Fraction in solids in dissolver: 0.022 (as AgI and PdI_2) [Vondra, 1977a,b]
- 0.011 of I is in PdI_2 which is assumed to decompose during melting of clad waste and solids and goes to off-gas system
- 0.011 of I in AgI which is assumed to be stable during melting of clad waste and becomes part of the clad waste

Fraction retained in dissolver solution: 0.0072 [Vondra, 1977b]. This is assumed to be volatilized in the vitrifier and goes to the off-gas system.

Ultimate end-point of iodine from reprocessing:
- 0.5% is released to the atmosphere (DF of 200 required by [EPA, 1977])
- 1.1% is incorporated into cladding waste in the form of AgI (see above)
- Difference (100%-0.5%-1.1% = 98.4%) is incorporated into an iodine waste form

^{14}C

Fraction assumed to be released to off-gas system from voloxidation and dissolution: 1.00

Fraction released to off-gas system that is assumed to be captured: 0.99

D. Cladding

Continue to support use of 0.05% of non-volatile SNF being associated with the cladding based on the following inconsistent information:

- Historical reports use this value [Kee, 1976; DOE, 1986]

- Statement that after repeated leaching of Zr cladding with boiling nitric acid the Pu content was reduced to 0.0005% [Blomeke, 1972]

- Information from May 2007 AREVA presentation to the Committee [ACNW&M, 2007] indicated 0.1% of Pu is in final waste forms (p 8 of presentation) and 0.04% of the alpha activity in the waste is in the cladding. This implies that 0.000004% of the SNF is associated with the cladding.

II. UREX Process Step

Fraction U assumed to be in U product: 0.997

Fraction Tc in U product: see earlier Tc discussion

Fraction of rare earth elements assumed to be in uranium: 0.0025

ASTM C 788 [ASTM, 2007] limits TRU alpha to 6.8 nCi/g U and Np to 3.4 nCi/g U. For Np this implies that 0.875% of the soluble Np follows the U stream. Allocating the remainder of the allowance to the limit (i.e., 6.8 - 3.4 = 3.4 nCi/g) for TRU elements other than Np yields a DF for Pu, Am, and Cm of 4.29E-07.

Fraction of other elements assumed to be in uranium product: 0.0

IV. CCD-PEG Process Step [Pereira, 2007]

Fraction of Cs, Sr, Ba, Ra, Rb, K, Na fed to CCD-PEG that is in product stream: 1.0

Fraction of rare earths fed to CCD-PEG that is in product stream: 0.0007

Fraction of other elements fed to CCD-PEG in product stream: 0.0

V. TRUEX Process Step [Chandler, 1956]

Fraction of rare earths in TRUEX feed going to TRU product: 0.0009

Fraction of U, Np, Pu in TRUEX feed going to TRU product: 0.999

Fraction of Am, Cm in TRUEX feed going to TRU product: 0.9997

Fraction of Th, Pa in TRUEX feed assumed to go to TRU product: 0.01

Fraction of Tc in TRUEX feed assumed to go to TRU product: see Tc discussion above

Fraction of other fission products in TRUEX feed assumed to go to TRU product: 0.00001

VI. <u>TALSPEAK Process Step [DOE, 1998; Barre, 2000; IAEA, 2005; Wymer, 1981]</u>

Note: Feed is TRUEX product, not TRUEX raffinate

Fraction of U and Np fed to TALSPEAK that goes to fission product waste: 0.001

Fraction of Pu fed to TALSPEAK that goes to fission product waste [AIChE, 1969]: 0.0001.

Fraction of Am and heavier that goes to fission product waste [TALSPEAK, 1999]: 0.0003

Fraction of Th and Pa assumed to go to fission product waste: 0.99

VII. <u>Notes on Final Output Stream Characteristics Assumptions</u>

A. <u>Volatile Effluents</u>

Not applicable; goes up the stack

B. <u>Tritium Volatile Waste</u>

Tritium is assumed to be made into tritiated water by catalytic conversion [IAEA, 2004] and incorporated into polymer-impregnated cement based on studies showing at least 10 times less leaching from polymer-impregnated concrete (PIC) [Albenesius,1983]

10 percent by weight of polymer replacing water [Blaga, 1985]

53 wt% water is optimal [PCA, 2007] although the ratio can range down to about 45 wt%. A larger value was used to account for higher density of water containing deuterium and tritium.

PIC grout density is 2.2 g/cc [Blaga, 1985]

Water density and tritium content

- Hydrogen in water made from dissolver offgas (spent nuclear fuel (SNF) water) is 84 wt% tritium, 1 wt% deuterium, and 15 wt-% hydrogen based on ORIGEN2 output (ML072820458) and ratios of fission product yields for hydrogen isotopes [IAEA, 2000].

- The average molecular weight of recovered hydrogen is 2.7 and average molecular weight of water made from the hydrogen is 21.4.

- The water density is 1.19 g/cc. Each gram of water contains 0.12 grams of tritium.

2.2 g cement contains 2.2x0.53 = 1.17 g normal water or 1.17x1.19 = 1.39 g SNF water or 1.39x0.12 = 0.17 g tritium. Thus, 0.076 g T/g cement

C. ^{14}C Volatile Waste

99 percent of the ^{14}C assumed to be recovered from the dissolver off-gas using molecular sieves and scrubbed with calcium hydroxide slurry to yield calcium carbonate [DOE, 1986].

Calcium carbonate is assumed to be fixed in grout [Croff, 1976].

- Grout density is 1.6 g/cc [Croff, 1976]
- Grout loading is 30 wt% [Croff, 1976]

Calcium carbonate is 12 wt% carbon

- Carbon is 0.08 wt% ^{14}C [DOE, 1986]
- Leads to 0.31 x 1.6 x 0.12 x 0.0008 = 4.6E-05 g ^{14}C /g waste

D. Krypton Volatile Waste

Kr is recovered using cryogenic distillation [DOE, 1986]

At 25 years of decay, there is 351g Kr/MTIHM (1.34 wt% Kr-85) and 5357 g/MTIHM xenon based on ORIGEN2 calculation

- Krypton recovery is 0.85x351 =298 g/MTIHM.
- The ratio of xenon in product to krypton in product ranges from 25 wt% [DOE, 1986] to 12.5 vol% (18 wt%) [IAEA, 1980]. Defer to IAEA value (18 wt%) that is based on pilot plant experience.

Assumed to be stored in compressed gas cylinders at 1.5 atmospheres (Barnwell LLW disposal site license condition limiting pressure) [DHEC, 2000].
- Ignore cylinder volume
- Kr load factor is 0.0134x(1-0.18) = 0.011 g Kr/g noble gas in cylinder
- Kr density in gas is 0.0047 g Kr/cc noble gas in cylinder at 1.5 atmospheres pressure

E. Iodine Volatile Waste

It is assumed that silver mordenite (AgZ) sorbent that is grouted contain 34 wt% AgZ. Use information in Table XI of [IAEA, 1987] for I loading as follows:

- Density of grouted AgZ is 2.1 g/cc [IAEA, 1987]
- From ORIGEN2 calculation iodine is 180 g ^{129}I /MTIHM and 236 g total iodine/MTIHM
- ^{129}I loading in grout is 625 kg I x (180 kg ^{129}I/kg total I)/(3470 kg AgZ/0.34 kg AgZ per kg waste) = 0.0414 g ^{129}I/g waste.

F. Cladding Waste plus Technetium, Dissolver Solids, and a Fraction of Nonvolatile-SNF

It is assumed that all cladding and other structural material (end pieces, grid spacers) will be melted into an alloy for disposal.

Recovered technetium, dissolver solids, and a fraction of non-volatile SNF are included. However, no tritium is included because ZrT_2 is assumed to be dissociated by voloxidation or melting (see earlier discussion on tritium).

Radionuclide density is 1.0 because the entire waste form is composed of waste materials.

Density is the mass-weighted average of Zr (for Zircaloy) and SS (for SS, Inconel, and Nicrobraze) which is 6.8 g/cc [Croff, 1978].

G. Uranium Product

Density of product can have a wide range because the degree of compaction is unknown, and the oxidation state is unknown; a value of 3.5 g/cc is used.

- UO_2 powder densities range from 2.0 to 5.9 [Croff, 2000]. However, product is unlikely to have a high dioxide concentration because of the cost of oxide reduction.
- U_3O_8 densities range from 1.5 to 4.0 [Croff, 2000].
- The product of the DOE de-fluorination plants is a mix of the two oxides with more U_3O_8 than UO_2. The higher end of the U_3O_8 density range is selected to account for the UO_2 component.

H. TRU Product

It is assumed that it is converted to an oxide (mainly dioxides) and fabricated into pellets.

The theoretical density of fuel is calculated based on literature values [Weast, 1968; Corliss, 1964] weighted by mass in ORIGEN2 TRU product

It is assumed that pellets are 95% of theoretical density

I. Cesium/strontium Waste
Assume Cesium/strontium is made into an aluminosilicate waste form using steam reforming

Bulk density of product is 1 g/cc [McGrail, 2003].

Waste loading is 27 percent [Jantzen, 2002].

J. Fission Product Waste

Values are based on experience at DWPF.
- Glass density is 2.65 g/cc [Bibler, 2000].
- Waste loading is 38 percent [Jantzen, 2004].

VIII. Summary of ORIGEN2 results

Table E1 contains the mass and radioactivity of selected spent fuel constituents taken from the ORIGEN2 calculation performed using the preceding input.

Table E1: Mass and Radioactivity of Selected Constituents of 25-Year-Old Spent Nuclear Fuel Irradiated to 33 GWd Per MTIHM

Constituent	Mass, g/MTIHM	Radioactivity, Ci/MTIHM
Dissolver solids	2.88E+03	2.00E+00
Tc	1.16E+02	1.97E+00
Ru	1.10E+03	1.000e-02
Pd	2.77E+02	2.300e-02
Mo	1.34E+03	0.00E+00
Rh	4.70E+01	2.000e-03
Tritium	2.080e-02 [39][a]	2.01E+02
Krypton	1.590e+00 [42.4]	1.85E+03
Xe	5.350e+00 [894]	0.00E+00
Iodine	2.36E+02	3.200e-02
I-127	5.59E+01	0.00E+00
I-129	1.80E+02	3.200e-02
Carbon	9.680e+01 [155,000]	6.000e-01
C-12	8.83E+01	0.00E+00
C-13	8.38E+00	0.00E+00
C-14	1.340e-01	6.000e-01
Cesium	2.11E+03	5.85E+04
Cs-133	1.13E+03	0.00E+00
Cs-134	2.600e-02	3.39E+01
Cs-135	3.01E+02	3.500e-01
Cs-137	6.71E+02	5.84E+04
Strontium	6.48E+02	4.01E+04
Sr-86	4.000e-01	0.00E+00
Sr-88	2.50E+02	0.00E+00
Sr-90	2.94E+02	4.01E+04
Uranium	9.56E+05	2.83E+00
U-232	1.300e-03	2.770e-02
U-233	5.000e-03	4.850e-05
U-234	2.06E+02	1.29E+00
U-235	7.98E+03	1.730e-02
U-236	3.97E+03	2.570e-01
U-237	1.130e-05	9.300e-01
U-238	9.44E+05	3.180e-01

Table E1 (Continued): <u>Mass and Radioactivity of Selected Constituents of 25-Year-Old Spent Nuclear Fuel Irradiated to 33 GWd Per Metric MTIHM</u>

Constituent	Mass, g/MTIHM	Radioactivity, Ci/MTIHM
Neptunium	4.63E+02	1.74E+01
Np-236	4.63E+02	3.300e-01
Np-237	4.120e-04	3.200e-02
Np-238	1.250e-07	1.70E+01
Plutonium	8.28E+03	4.06E+04
Pu-236	2.690e-06	1.400e-03
Pu-238	1.21E+04	2.07E+03
Pu-239	5.03E+03	3.11E+02
Pu-240	2.32E+03	5.28E+02
Pu-241	3.66E+02	3.77E+04
Pu-242	4.51E+02	1.72E+00
Pu-244	2.400e-02	4.220e-07
Americium	9.50E+02	3.00E+03
Am-241	8.64E+02	2.97E+03
Am-242	7.960e-06	6.44E+00
Am-242m	6.700e-01	6.47E+00
Am-243	8.55E+01	1.70E+01
Curium	1.04E+01	7.60E+02
Cm-242	1.600e-03	5.33E+00
Cm-243	2.200e-01	1.15E+01
Cm-244	9.18E+00	7.43E+02
Cm-245	8.500e-01	1.500e-02
Cm-246	1.000e-01	3.100e-03
Cm-247	9.100e-04	8.420e-08
Cm-248	4.440e-05	1.730e-07

[a]Volume in cubic centimeters at standard temperature and pressure

www.ingramcontent.com/pod-product-compliance
Lightning Source LLC
Chambersburg PA
CBHW080241180526
45167CB00006B/2363